5

Thinking by numbers

Written by TIM HEATH

Series editor STEVE HIGGINS

OXFORD
UNIVERSITY PRESS

OXFORD
UNIVERSITY PRESS

Great Clarendon Street, Oxford OX2 6DP

Oxford University Press is a department of the University of Oxford.
It furthers the University's objective of excellence in research,
scholarship, and education by publishing worldwide in

Oxford   New York

Auckland   Cape Town   Dar es Salaam   Hong Kong   Karachi
Kuala Lumpur   Madrid   Melbourne   Mexico City   Nairobi
New Delhi   Shanghai   Taipei   Toronto

With offices in

Argentina   Austria   Brazil   Chile   Czech Republic   France   Greece
Guatemala   Hungary   Italy   Japan   South Korea   Poland   Portugal
Singapore   Switzerland   Thailand   Turkey   Ukraine   Vietnam

Oxford is a registered trade mark of Oxford University Press
in the UK and in certain other countries

Activity text © Tim Heath 2005
Introduction text © Steve Higgins 2005

The moral rights of the author have been asserted

Database right Oxford University Press (maker)

First published 2005

British Library Cataloguing in Publication Data

Data available

ISBN-13: 9780198361275
ISBN-10: 019 836127 0

3 5 7 9 10 8 6 4 2

Illustrated by Simon Smith
Typeset in Great Britain by Artistix, Thame, Oxon
Printed in Great Britain by Ashford Colour Press, Gosport, Hants

Contents

Introduction

Thinking by Numbers aims to develop thinking skills through mathematics lessons and activities across the primary age range. Although it can be used by an individual teacher, we think that you will get the best from the series if you use the activities across your school to undertake a professional inquiry into the potential of these lessons to develop pupils' thinking. Hence, the sections on *Professional development* (page 9), *Classroom management* (page 12), *Formative assessment and assessment for learning* (page 14), and *Speaking and listening* (page 18) are important aspects of the series. These sections will support you in helping to make the activities successful, as well as suggesting opportunities to develop aspects of your own teaching. Most of these introductory sections also contain suggestions for further reading that will support your exploration of thinking skills through the activities in *Thinking by Numbers*.

Teaching children to think for themselves is at the heart of primary education. It is all too easy to focus on the demands of the curriculum and its assessment and forget that the facts and knowledge have to be connected with an understanding of this curriculum content to help the learner make sense of it all. Without this understanding learners cannot use the information they have been taught and see how it relates to other ideas or knowledge that they have already. At the core of the thinking skills movement in education is the belief that this kind of thinking is teachable. This belief has been inspired by the work of two leading educators.

History of thinking skills

In Israel after the Second World War, many refugee children had been through traumatic early experiences. On traditional tests, such as IQ tests or standardized tests of achievement, many of these children scored so badly that they seemed 'unteachable'. Working to integrate such children Reuven Feuerstein refused to accept this conclusion and devised ways to find out exactly which kinds of thinking they were unable to do, how they could be helped to develop these skills, and, therefore, each individual's *potential* for learning.

Feuerstein developed a set of techniques and tasks called 'instruments' that helped these learners succeed on subsequent tests. These methods were termed 'dynamic', in the sense that children were studying the process of learning and the change that took place. Feuerstein argued that such a process was much more likely to predict how a person might then learn in the future. Many of Feuerstein's ideas have influenced work on teaching thinking skills, in particular his emphasis on the importance of the interaction of the teacher, or 'mediation' of thinking.

Another important figure in thinking skills (or 'Critical Thinking', as it is called in the United States), is the American philosopher Matthew Lipman. As a university professor, he thought that his students had been encouraged to learn facts and to accept opinions, but not to think for themselves. He developed a programme, therefore, called 'Philosophy for Children', which aims to help younger people (from six-year-olds to teenagers) to think by raising questions about stories that they read together. The teacher uses children's natural curiosity about the stories in order to promote active participation and learning. One of Lipman's basic convictions is that children are natural philosophers, and that they view the world around them with curiosity and wonder, which can be used as a basis for thinking and reasoning.

Both Feuerstein and Lipman, though from very different starting-points, hold a similar belief in children's abilities. They have demonstrated that through thinking exercises and activities learners can exceed the predictions of achievement which tests may have suggested is their limit of competence. This, then, forms the basis of techniques in thinking skills – realizing children's potential. Their work has inspired many others to explore and develop approaches which help children to become more effective learners as they start to think for themselves. The aim of this book is to help you, as a teacher, to see how this kind of thinking can be developed.

Teaching thinking

Some people argue that the idea of trying to teach general thinking skills is misguided because in practice thinking always occurs in a specific situation. Further, they believe that it is better to concentrate on teaching subjects and developing specific and detailed knowledge. However, *Thinking by Numbers* has been developed on the principle that there are common features of thinking in different situations, that it is helpful to try to apply techniques learned previously in new situations. For example, once you have used a graphic organizer, such as a Venn diagram, to compare and contrast themes in traditional tales in literacy, you can use the same technique to compare and contrast in other curriculum areas, such as family life in different eras in history.

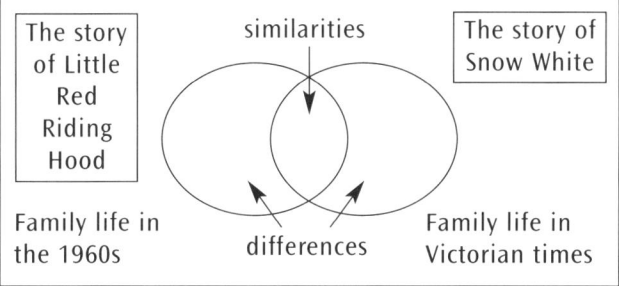

Since 1999, the national curricula for England and Wales now specifically include thinking skills (see page 6 for more details). In Scotland the *5–14 Guidelines* emphasize the capacity for independent thought through enquiry, problem solving, information handling and reasoning, as well as identifying learning and thinking skills in the core skills and capabilities. So the current challenge for teachers is not whether to teach thinking skills, but how best to teach them!

Approaches to teaching thinking skills

There is a host of different programmes and approaches which advocate teaching thinking. These can be categorized broadly into whether they adopt an 'enrichment' approach where they are taught through extra or separate lessons, or an 'infusion' approach where the particular skills are taught through the normal lessons that schools provide. There are certain advantages and disadvantages to adopting these two different teaching approaches. If thinking skills are taught separately it is possible to make skills and techniques explicit, but there is a danger that they may not be used, except in special 'thinking' lessons. However, if they are taught as part of other lessons, such as mathematics or history, there is also a danger that the skills and techniques will become submerged by the curriculum content and not be seen as skills that can be applied elsewhere.

We believe that it is necessary to do both – to have a mixture of 'thinking lessons' with discussion of the kinds of thinking that are involved, and 'subject lessons' where skills can be applied and developed, but perhaps less explicitly. Identifying some lessons as 'thinking maths' lessons gives a clear signal to the children that you are looking for something different in the way that they work and the way they talk and listen. It is challenging to make the time to develop speculation or reasoning in every lesson, but it is also difficult to make sure it happens at *some* time in *some* lessons. We suggest that the activities in the different units can be used as a way to emphasize aspects of thinking that you wish to develop. You may then choose to develop other similar lessons where you can re-use the structure of the activities, or use some of the ideas and techniques in other subject areas.

> ### *Suggestions for further reading*
> H. Sharron and M. Coulter, *Changing Children's Minds: Feuerstein's Revolution in the teaching of Intelligence* (Birmingham, Questions Publishing Company, 1994)
>
> M. Lipman, *Thinking in Education* (Cambridge University Press, 2003)
>
> C. McGuinness, *From thinking skills to thinking classrooms: A review and evaluation of approaches for developing pupils' thinking* [DfEE Research Report RR115] (Norwich, HMSO, 1999)
>
> V. Wilson, *Can Thinking Skills Be Taught? A paper for discussion* (Edinburgh, Scottish Council for Research in Education, 2000) [Available at: http://www.scre.ac.uk/scot-research/thinking/index.html]

Thinking skills and the National Curriculum

Classifying thinking skills

There are more ways to think about thinking than you could imagine! Amongst the wealth of lists, frameworks, models and taxonomies of thinking that have been developed, many people have heard of the 'Bloom's Taxonomy', which is considered the original way of classifying 'higher order thinking'. This taxonomy is basically a three-tier model:

- **knowledge** – in the form of facts, concepts, rules or skills
- **basic thinking** – relatively simple ways of understanding, elaborating and using what is known
- **higher order thinking** – a learning process which leads to a deeper understanding of the nature, justification, implications and value of what is known.

The National Curriculum in England uses five classification headings to denote thinking skills that should be embedded across all subject areas so pupils learn how to learn. These are:

- evaluation
- creativity
- enquiry
- reasoning
- information processing.

However, no single classification or framework can ever fully describe the complexity of all the kinds of thinking we experience. What is missing in both the original version of Bloom's work and in the National Curriculum list is the role of the thinker in thinking – one's own awareness, reflection and engagement. This metacognitive component (i.e. thinking about how we think) is an essential ingredient in developing a learner's understanding of their own thinking and the ability to think for oneself. The following table shows how *Thinking by Numbers* works alongside these thinking classifications to develop thinking skills.

Bloom's Taxonomy	National Curriculum	Thinking by Numbers	
Knowledge Abstracts and universals Using specifics Knowledge of specifics	*Information processing*	Unit 1: Sort it out!	*Unit 6: Think on!*
Basic thinking Application Comprehension			
Higher order thinking Evaluation Synthesis Analysis	*Reasoning* *Enquiry* *Creativity* *Evaluation*	Unit 2: That's because … Unit 3: Detective work Unit 4: What if? Unit 5: In my opinion	

Comparison of Bloom's Taxonomy, the National Curriculum and *Thinking by Numbers*

The National Curriculum

The National Curriculum categories contain the following breakdown of skills, which form the basis for the units in the *Thinking by Numbers* series.

Information processing skills

These enable pupils to locate and collect relevant information, to sort, classify, sequence, compare and contrast, and to analyse part/whole relationships.

Reasoning skills

These enable pupils to give reasons for opinions and actions, to draw inferences and make deductions, to use precise language to explain what they think, and to make judgements and decisions informed by reasons or evidence.

Enquiry skills

These enable pupils to ask relevant questions, to pose and define problems, to plan what to do and how to research, to predict outcomes and anticipate consequences, and to test conclusions and improve ideas.

Creative thinking skills

These enable pupils to generate and extend ideas, to suggest hypotheses, to apply imagination, and to look for alternative innovative outcomes.

Evaluation skills

These enable pupils to evaluate information, to judge the value of what they read, hear and do, to develop criteria for judging the value of their own and others' work or ideas, and to have confidence in their judgements.

The first five units of the *Thinking by Numbers* books are based on these classifications. We have also added two further components:

- ❯ a final unit which provides opportunities for **using and applying thinking skills** covered in the earlier units
- ❯ a **metacognitive skills** element running throughout all of the activities, which aims to develop children's awareness and understanding of the thinking they are doing.

Suggestions for further reading
L.W. Anderson and D.R. Krathwohl (eds.), *A Taxonomy for Learning, Teaching and Assessing: A revision of Bloom's Taxonomy of Educational Objectives* (New York, Longman, 2001)

S. Higgins, J. Miller, D. Moseley and J. Elliot, 'Taxonomy Heaven', *Teaching Thinking, 12,* (Autumn 2003)

Thinking skills in mathematics

Mathematics is an area of the curriculum which is full of opportunities to develop pupils' thinking skills and reasoning abilities. An emphasis on developing strategies, identifying patterns and rules, and clarifying concepts helps children learn mathematics by making aspects of it more explicit in the classroom. Developing reasoning, problem solving and enquiry skills through mathematics can support the development of these 'higher order' thinking skills more widely, and encourage successful learning in other subjects. A number of principles underpin the activities in each of the six units in *Thinking by Numbers*. These will help pupils to see the connections between the way that they have worked on a mathematics task and then how they can apply these skills in other contexts, either in other areas of mathematics or other areas of learning and understanding.

Challenge

Thinking activities must provide a level of challenge. This means that they should not be too easy to complete, nor so hard that the pupils cannot recognize that they have been successful. Alternatively, the activities may have more than one solution, or route to a solution, that can be evaluated by the pupils to decide which is the best answer or approach. Mathematics is a subject which people often think they just 'can't do'. Successfully completing challenges encourages pupils to see that maths is a subject that they can learn to be good at.

Active discussion

Thinking activities need to be talked about. Mathematics has both a vocabulary and a language of its own. Familiar words are used in unfamiliar ways, such as 'product' or 'difference', and it has its own terminology, such as 'numerator' or 'perpendicular'. Pupils will need time to practise speaking mathematically and explain what they are thinking using this language. This can be difficult to do with the whole class, so some paired or small group work is essential to provide opportunities to explore ideas and allow pupils to develop confidence with the vocabulary.

Feedback

Giving feedback will be key in ensuring pupils make progress in a thinking activity. One of the easiest ways to do this is to have 'mini-plenaries' as the lesson develops. Stop the class for a few minutes and ask a group to explain where they are up to. This will give you the chance to highlight successful ways of working, as well as asking for reasons and challenging their thinking.

Review

When developing thinking skills it is important to review both the **content** of the activity and the **process** that the pupils have used to complete the activity. This means talking about the mathematics involved in the task and the way that they have worked (the skills used in collaborating, working systematically, or identifying patterns and rules). It is often helpful to discuss the latter the next time pupils undertake a similar task so that you can remind them of what was successful. A combination of 'mini-plenaries' throughout the lesson, a review at the end of a lesson, then recapping at the beginning of the next lesson will help ensure children understand that you want them to think not just about *what* they have learned, but *how* they have learned it.

Professional development

We advocate that you try out the *Thinking by Numbers* activities as part of your professional development programme. A critical perspective on the lesson is essential. The activities alone will not succeed in developing thinking skills without this perspective. It is helpful to have a colleague with whom to discuss the activities as you try out the different ideas. We believe that a key part of teaching thinking and thinking skills successfully is to have some time and space to reflect on your own teaching so as to increase the emphasis on developing pupils' understanding. The introductory 'brief' and final 'debrief' sections of each activity aim to support this by summarizing the key features of the lessons and indicating aspects for review.

Thinking by Numbers provides a combination of teacher-led activities, then discussion and collaborative working in small groups, followed by some kind of whole class discussion, or plenary, which reviews both the content and process of learning. The results of this approach are usually a higher level of engagement in the activities, more talking and discussion about the activities. The activities themselves are open-ended to the extent that genuine discussion is not only possible but helpful. They are also challenging but enjoyable activities, helping to create a classroom climate where there is an emphasis on succeeding after effort.

As part of this process you should get more opportunities to hear what your pupils think. As you plan these lessons to increase engagement in learning you will need to listen carefully to how your pupils respond. The enjoyment should initially help to sustain more permanent changes in patterns of classroom interaction. The further feedback you get from insights into pupils' understanding will help identify any misunderstanding or misconceptions that you can tackle through 'mediation' or questioning and discussion.

Some suggestions for getting started:

1   **Work with a colleague**. This might be a colleague teaching the same year group, in which case you can investigate the impact of the same activities. Alternatively you may be working with a colleague in another year group, so you might look at similar kinds of activities or similar aspects of thinking. Working with a colleague means you are more likely to keep to your plan, building progress in time for review. Discussing things with someone else helps to clarify our own thinking, and makes it easier to see patterns or themes in what has happened.

2   **Decide what you want to investigate or improve**. It is easier to develop children's thinking if you focus on a particular area that you feel needs improvement. You could:
    - identify information processing as a key mathematical skill needing improvement
    - focus on your own questioning and how you probe and challenge your children's thinking
    - develop more precise use of mathematical language
    - aim to increase participation in lessons by children who are not usually engaged.

3   **Set a timescale** (at least eight weeks, up to a school year) and plan which activities you are going to use. How often will you have *Thinking by Numbers* lessons? Once a week? Once a fortnight? How will you make sure you have time to review the activities with a colleague?

4   **Try out the activities and review** them as soon afterwards as you can with your colleague. What was different in the lesson compared with other maths lessons? Were you able to see patterns in the children's thinking? Were there any common misconceptions that you needed to tackle? How well did the collaborative tasks go?

5   **Analyse what happened**. If there is improvement, what do you think caused it? The focused practice? Your extra time and effort? The pupils' discussion? Your understanding of their thinking? Would it probably have happened anyway?

6   **Review progress**. What have you learned that you can apply in the longer term? Do some kinds of questions work better than others? Can you use any of the strategies more widely?

How to use *Thinking by Numbers*

The activities in *Thinking by Numbers* can be used in different ways – there is no need to work through them in order, though the final unit is designed to let pupils apply the skills that they have developed. Therefore, for you to assess how well these skills have been learned, it should be used after some of the other activities in the first five units. Some of the activities are based on thinking skills strategies which can be used more widely either in mathematics or other subjects of the curriculum. You should therefore evaluate if there are any aspects of the activity or teaching technique which could be used more generally. Although the books are aimed at different age groups, you may find activities that you can use or adapt in other books in the series. This is particularly true of the generic strategies, such as 'odd one out', which can be used again in mathematics or other areas of the curriculum. See page 24 for a fuller description of some generic activities used throughout the series.

Just as enquiry is at the heart of thinking skills activities for pupils, we believe that it also needs to be a part of the way you use them as a teacher. None of the activities will work by themselves, and they will not all be equally effective since this depends on the existing skills and knowledge of your pupils. You will have to use them critically to see how they can help your pupils' thinking – it is impossible to do this directly, since we cannot see into our pupils' heads and know what they are thinking. Nevertheless, it is possible to plan a series of activities that enable you to find out about pupils' thinking at different times and in different ways. This allows you to infer their level of understanding. Therefore, this needs to be a process of enquiry – finding out what and how your pupils think. The 'Watch out for' and 'Listen for' sections of each activity should help with this process.

The units

The units are based around the classification of thinking skills in the National Curriculum for England and the headings of **information processing**, **reasoning**, **enquiry**, **creative thinking** and **evaluation**. Each unit begins with an overview of these particular aspects of thinking, and ends with a summary looking at how these skills can be developed. Of course, it is not possible to separate the thinking in different activities so that they only involve reasoning or creative thinking. Thinking is a complex activity which involves all kinds of thinking at the same time. It is holistic, multi-dimensional and dependent upon the context that we find ourselves in. The purpose of the tasks in *Thinking by Numbers* is to enable you to focus on a particular kind of thinking and to consider how it can be developed or fostered in your pupils.

Links

The appendices (pages 88 to 95) contain information about how *Thinking by Numbers* relates to the *Framework for Teaching Mathematics* used in England, and the *5–14 Guidelines* for Scotland. A glossary of thinking skills terms is also included on page 96 for reference.

Suggestions for further reading
P. Adey, *The Professional Development of Teachers: Practice and Theory* (Dordrecht, Kluwer Wolters, 2004)

S. Higgins, *Thinking Through Primary Teaching* (Cambridge, Chris Kington Publishing, 2001)

The activities

4 Each activity has a whole class introduction where you will be 'Setting the Scene' and modelling the problem to the children.

1 Each activity has an introduction, 'The Brief', and review points, 'The Debrief', to explain the context for the activity. This is the 'professional development' part to help you consider what you want to achieve in the thinking lesson, and later to review how well it achieved its thinking skills aims.

6 The 'Checkpoints' section gives ideas for how to the keep the activity on track. This section also has suggestions on what to watch and listen out for, and prompts and pointers to stimulate discussion.

Easy as pi

BRIEF

In this investigation the children gather information and find a pattern in the data that they collect. Groups must organize themselves well and ensure that everyone has a part to play. The children decide how much information they need to discover the anticipated pattern. Ingenuity in devising and using reliable measuring techniques will prove helpful.

Key maths links
- Measures
- Problems involving measures

Thinking skills
- Information processing skills
- Planning what to do
- Comparing

Language
length, width, perimeter, centimetre, millimetre, ruler, tape measure

Resources
PCM 3 (one per group)
PCM 4 (as needed)
three or four **large paper squares of different sizes**
circular objects of various sizes (bottles, tins and other containers)
measuring equipment: tape measures, rulers, string, calipers etc.
calculators

Setting the scene

Display three or four large squares of various sizes cut from coloured paper (or draw them on a board or flipchart). Explain that you want to discover if the dimensions of each square have anything in common. Using a tape measure, or length of string and ruler, measure the perimeter of one square and note this on the shape, (N.B. do not calculate the perimeter by multiplying the edge length by 4). Record the width of the shape. Repeat for the other squares. *Do these measurements have anything in common?* Discuss. Draw a table to summarize the measurements:

Perimeter of square	Width of square	Linking number (perimeter ÷ width)
84 cm	21 cm	4

Calculate the value of 'perimeter ÷ width' for each shape. Call this the linking number. Of course, within the limits of your measurements the result will be 4 in each case. The children will be able to explain this result. *Would something similar happen with circles?*

Getting started

Supply a selection of objects with a circular cross-section. Ask the children to work in small groups to undertake an investigation of the link between the 'distance around a circle' and its 'width'. Allow the children to plan and organize themselves. Let them make mistakes.

Simplify
If you judge that the children need guidance they can use PCM 1a to record results. Part of the challenge of this activity is to find ways to gather reliable information. However, if you judge children are struggling this is an ideal opportunity to teach specific techniques, e.g. using calipers or parallel straight edges for the circle's width, and wrapping string or a paper strip once or more around the object to measure its circumference (PCM 1b).

Challenge
If the children are skilled at making line graphs they can record measurements graphically (use PCM 2: the dashed lines indicate linking ratios of 3 and 4 and plotted results are likely to lie between the two). Repeat measurements and calculate average values to improve accuracy. You could use appropriate technical vocabulary: *circumference, diameter, ratio*. However do not obscure the point of the activity by introducing too much that is new.

Checkpoints

You will want to discuss the accuracy of the children's measurements. Acknowledge that 100% accuracy is impossible but emphasize the need for techniques that supply reliable results. When the children calculate the ratio linking the distance around each circle with its width they will find their calculators will display numbers with many decimal places. Consider how to handle these values with the children.

! Watch out for ...
Some groups will be tempted to try to measure all the objects. If haste compromises accuracy then the children's results will be unreliable. Ensure the children work as carefully as possible. Encourage all of them to take a part in the practical work.

? Ask ...
- How many measurements will you take?
- What do you expect?
- Is there a link? Are you sure?

" Listen for ...
Applaud the children's efforts to gather accurate information: *I'm going to check that measurement.*

Moving on ...

Compare the outcome of different groups' investigations. Do the results indicate a consistent link between the perimeter of a circle and its width? What is the 'linking number' for circles? Is there much variation in its calculated value? Why? (You could explain that the linking number for circles is called 'pi' (π) and give a little more information about it. However, calculations using π are still some years away for most children and will almost certainly not be worth attempting.) Also consider how each child contributed to the investigation. How did groups organize themselves to complete the task? Did everyone have a part to play?

Where next?
- Extend the investigation to other shapes. What is the linking number for a range of regular polygons?

DEBRIEF

Did you give the children enough freedom to direct their own investigation? Were they able to exercise ingenuity in gathering information? Did they organize their data successfully and make the looked-for link? Did groups make mistakes? And learn from them?

2 Basic information about mathematical objectives and language are included, along with any resources needed, plus the thinking skills focus.

5 The 'Getting started' section shows how the activity can be developed through collaborative group work.

7 Reviewing progress and stimulating further thinking are covered in the 'Moving on' section, as are suggestions to develop the teaching strategy or approach in other mathematics lessons, or in other subjects in 'Where next?'.

3 Each activity has accompanying photocopiable resources. Some are resource sheets for the activities, others aim to support the recording of the activities, particularly by pairs or small groups of pupils (for more information on recording see page 16).

Classroom management

Structure and timing of lessons

Each book in the *Thinking by Numbers* series comprises six units. These are based around the English National Curriculum thinking skills headings of **information processing, reasoning, enquiry, creative thinking** and **evaluation**, with a final unit focused on using and applying the skills acquired through the earlier activities. Each unit contains two activities, with three in the final 'using and applying' unit, giving a total of 13 thinking activities or lessons for each year group. They have been planned as mathematics lessons and cover aspects of the curriculum appropriate for each age group (see the NNS and *Mathematics 5–14* matching charts on pages 90 to 95). You could also use the activities as thinking lessons and follow the suggestions at the end of each unit to develop the ideas and thinking strategies across the curriculum. When planning how to use the activities there are a number of different approaches you could take, and these are outlined here.

Regular thinking skills development

You could work through the activities using a *Thinking by Numbers* activity every two or three weeks. The benefit of this approach is that it provides regular opportunities to highlight the thinking skills you want to develop across the year. In the intervening time you would need to make sure that you refer back to the lessons and activities, as it would be all too easy for your pupils, particularly younger children, to forget what you are looking for in their work in the thinking activities.

Intensive thinking skills development

You could choose to work through the units more intensively, perhaps one activity each week over a term, so that you could then take the skills and ideas further over the course of the year. This may also be more suitable for year groups in England where your teaching is affected by statutory tests, such as Year 6 in particular. A further advantage of this approach is that you can build up some momentum with regular 'thinking maths' lessons. Assuming they go well initially, the children will start to look forward to the lessons and you can then capitalize on this enthusiasm. You will also develop a language around the lessons and activities with your class, and the regular practice will enhance this development.

Integrated thinking skills development

Another possible approach for teachers in England and Scotland is to use the matching charts to the *NNS Framework* or *Mathematics 5–14*. These are provided in the appendices and will enable you to substitute the *Thinking by Numbers* activities where they fit most appropriately in your usual teaching plan. Whilst this is less disruptive to the mathematics curriculum, you will need to work hard to develop the thinking themes in the book. The thinking skills issue here is how you get the children to use what they learn elsewhere. This is always a challenge with any learning at school: how do you get learners to transfer what they know or can do to a new situation? The concept of 'bridging' is a useful one. As a teacher you connect or 'bridge' the knowledge or skills between different contexts. Where you have regular lessons you can mention things that you then refer to in other lessons. The further apart the sessions, the harder you will have to work to make those connections meaningful. *You remember when we used a Venn diagram to look at similarities and differences? Could we do something similar here?*

Managing the lesson

To use the *Thinking by Numbers* activities effectively you will need to think through the method of working. Your pupils will need to have a clear idea of what they are doing, and why, so that in the review sections of the lesson they can evaluate how successful they have been. It is important to get the lessons off to a good start, so the children will need a 'hook' or some initial stimulus to launch into the activity well. This can be either through the way you introduce the activity, the resources that are used, or perhaps the way you make it meaningful to the pupils, tapping into their particular interests or enthusiasms. It is hard to predict exactly how long the different activities will take to complete. Sometimes children become particularly enthusiastic about a particular task and you will struggle to get through everything that is suggested. On other occasions you will have time to review the activities and ask the children to reflect on their learning.

the opportunities to develop speaking and listening skills are outlined on pages 18 and 19.

Introducing the lesson

Each activity begins with some kind of whole class introduction or demonstration. In this part of the lesson it is important to explain the activity and its purpose clearly. You should make objectives explicit; explain what you want from the pupils in terms of how they should work and the kind of language they should use. You will need to get feedback from the pupils to evaluate whether they understand what they are doing and know what they will have to do in the next phase of the activity. You may also need to adapt the activities according to the needs of your pupils. Although the activities have been designed for particular ages of pupils, you will need to judge whether some alteration is needed to provide the appropriate level of challenge for your class.

During the lesson

In most of the activities the pupils apply or extend the ideas presented in the introduction by working collaboratively in pairs or small groups. When moving from whole-class to paired or group work, it is useful to discuss or mention how the pairs or groups are going to work together and what you are looking for. At the transition it is worth praising specific behaviours: *I liked the way you sorted out the number cards for your group, David.* Though it is also important to tailor this praise, particularly for older pupils who should be aware of supportive behaviours and active listening strategies: *Your group got started really quickly, Emma, what was it that you each did?* Reinforce the method of sharing ideas, explaining that they can do better together than they can separately, and that copying and ownership of ideas are not factors. The tasks themselves are designed to be challenging and to benefit from some discussion in small groups so that pupils don't just make up their minds quickly. The activities also contain suggestions for differentiation, with advice on simplifications and challenges that should help you to ensure that the level of challenge is maintained as the pupils work through the tasks. Further advice on

Reviewing the lesson

The hardest part of these activities is in helping pupils to see that particular tactics, strategies or approaches are helpful, without teaching specific solutions or answers. This will require some skilful questioning and discussion. It is important to review both the process that the pupils have used, particularly the collaborative skills of speaking and listening, as well as reviewing the curriculum content and knowledge and understanding of the activities.

It is also a good idea to review some of this as the lesson unfolds, rather than waiting until the end. Whilst the plenary seems to be the logical place to review the lesson, the pupils also know that the lesson is drawing to an end and it can be hard to maintain their interest. Mini-plenaries are, therefore, an essential teaching strategy which can help make the activities successful. These can be very brief, just checking where groups are up to, or sharing a successful technique or tactic being used by some children. *I noticed you've sorted the cards into different groups, can you tell the class how they are organized?* It boosts their confidence if you draw this to the attention of the whole class and gives other pupils who may not be on track a clear hint about what they could do.

Another possibility is to recap at the beginning of the next lesson. This is essential if the *Thinking by Numbers* sessions are a week or more apart. You need to remind the children that these are different lessons which require thinking, explaining, reasoning and evaluating. There should be more time for discussion about what went well previously and what skills or strategies they might find useful. The main aim is to help the pupils understand that they might not be able to see a solution immediately, but by thinking and working together they will be able to complete the activity successfully. In mathematics this is particularly important as it is a subject which pupils tend to think that they are either good at or not good at, rather than a subject that they can all learn to be better at!

Formative assessment and assessment for learning

Formative assessment is about intervening during teaching to improve learning. As a teacher you gather feedback about what is going on (either within a lesson or between lessons) and use that information to alter what you do subsequently. Assessment for learning is a more interactive approach that takes assessment a stage further by involving the learners in understanding what the specific learning objectives are for each activity/task/lesson so that they can judge how successful they have been in achieving them. This helps teachers and pupils to understand the criteria for being successful at learning, both for short term objectives as well as longer term goals about 'learning to learn' more effectively.

When assessing for learning it is important to give pupils feedback about what they can do to improve (rather than giving marks or feedback that simply indicates whether they are correct or not). One common technique is to get pupils to give you feedback about how well they think they are doing on an activity or a piece of work. This can be a simple thumbs up/down signal from the class, or getting pupils to use traffic light colours to self-assess a piece of work they have done – green for go ('I understand it and can go on'), orange for getting there ('I could do with a little bit of help'), red for stop ('I'm stuck').

Thinking skills approaches also involve formative assessment. Most of the activities are about giving you, the teacher, information about children's thinking. This lets you assess their understanding and make decisions about how to support the development of that thinking. In addition, pupils are expected to talk about their thinking as they undertake the tasks. Developing this metacognitive talk (talk about their own thinking) is a powerful technique which helps learners understand their learning better.

Furthermore, focusing on what makes for successful learning encourages judgement about that learning and moves the discussion away from the products or outputs (such as a complete page of calculations) to what has been learned (such as, 'I am finding subtraction more difficult than addition'). The concept of transfer is crucial here since it moves learning away from the particular to the more general. *What have you learned today that you can use in the future? What have you learned previously that will help you now?*

Both assessment for learning and thinking skills approaches use collaborative techniques for learning: paired and group work so that learners benefit from discussion with their peers. Both approaches highlight the role of the teacher in effective questioning and discussions with the pupils to move their thinking on. Assessment for learning and thinking skills approaches are clearly complementary. If you are developing formative assessment you will be developing children's thinking skills. If you are developing children's thinking skills and being explicit about the thinking they are doing with them, then this is formative assessment!

Suggestions for further reading

Primary National Strategy, *Excellence and Enjoyment: learning and teaching in the primary years. Planning and assessment for learning: assessment for learning* (Document code: DfES 0521-2004 G) (2004)

Assessment Reform Group, *Assessment for Learning: 10 principles* (London, QCA, 2002) (available online at: http://www.qca.org.uk/ages3-14/downloads/afl_principles.pdf)

P. Black, C. Harrison, C. Lee, B. Marshall and D. William, *Assessment for Learning. Putting it into practice.* (Maidenhead, Open University Press, 2003)

S. Clarke, *Unlocking Formative Assessment: Practical strategies for enhancing pupils' learning in the primary classroom* (London, Hodder and Stoughton, 2001)

How do you know it is working?

One of the greatest challenges in developing learners' thinking is assessing how well the activities are going. You should feel that the tasks and activities are giving the children opportunities to think and you should get direct and indirect evidence of this. There are a number of ways that you can start to gauge the impact of the activities.

Enjoyment

First and foremost the activities should be enjoyable, both for you and your class. It is important that the activities are regarded as fun because this helps the children to develop their confidence to discuss what they think. It encourages the children to offer opinions and ideas without the worry of being 'wrong'. This aspect of the activities is vital to ensure their success. Thinking is hard work, so it needs to be as enjoyable as possible!

Participation

Enjoyment should lead to increased engagement and involvement in the lessons. One of the ways that you can assess this is by keeping track of who participates. Are the contributions coming from those who are usually involved and usually speak in whole class discussions? Can you use the paired or group work to build pupils' confidence in contributing to a whole class discussion? *I thought that your suggestion was a really good one – can you explain it to the class?* Are you getting spontaneous contributions from those you normally have to ask directly?

Language

The next thing to watch for is language that indicates thinking and reasoning. Are the pupils giving reasons? Do they use words like *then, so, because*? Are they being tentative (*I think … It could be … It might be …*) or speculative (*What if …? How about if we …?*)? You can start the lesson by saying you want to hear particular phrases, and giving suggestions for how they may be used. Then you need to look out for these first when the children are working in pairs or small groups. Then encourage the children to give longer responses in class discussions, ask them for reasons or examples, or to comment on each other's ideas. One of the most effective ways of encouraging this is simply to wait longer when you ask a question, and wait a little bit longer at the end of the response whilst indicating that you want them to continue. In mathematics you should also see the children using specific vocabulary more precisely; for example, are they getting more accurate in the use of words like *number, numeral* and *digit*? Or terms like *side, corner, edge* and *vertex*? You should also pay attention to the questions that the children ask. If the lessons are successful, the children will be asking questions about the content of the learning (rather than just about what they have to do).

Reflection

If the activities are working the children should know that they have been successful and that they have been thinking hard. They should show growing awareness of this and be able to talk about their thinking. At first this will come out during the activities or just as you finish. It is a good tactic to get them to review and reflect at the beginning of the next *Thinking by Numbers* task; this will help remind them of what is expected in the next task as well as giving you a chance to assess how much they recall from last time!

Transfer

The long term goal of *Thinking by Numbers* is to develop transferable skills. Evidence of this is shown when children start to refer back to thinking skills activities in terms of what they have learned. You should, therefore, begin to notice that they are using and talking about the skills that they are developing in other maths lessons or in other subjects. If this is spontaneous or unprompted you know that they are using the thinking skills for themselves.

Recording

Opportunities for recording are identified in most of the activities. However, there are a number of issues you will need to consider. The activities are about developing thinking and the lesson must focus on this as the most important outcome. Recording can distract from this if the children become concerned with making sure they 'get it right' when they have to write something down. There are two main aspects of recording. The first is the recording of the particular task. Some of the photocopiable resources are explicitly designed for this. For other activities the children will need to think about the best way to record their thinking and their progress through the activity. The activities are often collaborative so you may need to make copies of the completed sheet for all the children in the group.

The second aspect of recording is to support review of the activities. The 'What did you learn today?' photocopiable sheet (see page 17) is designed to help with this. It may not be appropriate to use it for every activity, but it will help you review aspects of the lesson that enable the children to develop an understanding of their thinking and their learning (see *Formative assessment and assessment for learning* on page 14 for more information about developing thinking about learning). This aspect is cumulative and progressive as you will need to encourage the children to think about:

- their learning
- what they did
- what kind of thinking was involved
- how they worked together
- what lessons or skills they have learned that they can use in the future.

When planning how to incorporate recording into a thinking lesson, it is helpful to consider the following principles.

1 Recording should be purposeful
The record should either help with the process of the task or capture aspects of the thinking that it will be helpful to review.

2 Recording should be integral
If keeping track of what they are doing is not part of the task, it becomes an extra burden and less likely to be completed effectively.

3 Recording should be used
If you ask the children to make some notes on their thinking, or to use the 'What did you learn today?' PCM, you need to make use of it in a discussion either in that lesson or as part of setting the scene for the next activity.

4 Recording should be short
The lessons are about thinking and this needs to be the most important part of the lesson. You will not be able to capture everything that happens; you may need to have some kind of record to keep track of what has happened, but keep it as simple as possible.

What did you learn today?

Name _____ Date _____

> **What did you learn today?** _____
> _____
> _____

What kind of thinking did you do today?

	Yes	No
I remembered things that were useful	☐	☐
I organized my ideas	☐	☐
I thought of reasons why	☐	☐
I found out something I did not know	☐	☐
I used a rule or a pattern to work something out	☐	☐
I had a new idea which was helpful	☐	☐
I was methodical	☐	☐

How challenging was it?

Circle one of the choices on the line.

Very easy   Easy   OK   Hard   Very hard

Working with others

	Yes	No
I asked my teacher a good question	☐	☐
I asked a question which helped my partner	☐	☐
I asked a question which helped my group	☐	☐
I shared my ideas	☐	☐
I changed my mind after listening to someone	☐	☐
I was good at listening to my partner	☐	☐

Speaking and listening

Talking, thinking and learning are all closely related. We can remember things that we have heard, but it is only when we can put these ideas into our own words that we know we have learned them effectively. Speaking and listening are, therefore, at the heart of any thinking skills work. Listening to your pupils talk is also the best feedback you can get to assess what they are actually learning. It is therefore essential that the lessons and activities have speaking and listening at their core.

Children should be able to explain not just what they are doing, but why, and that their thinking is about the learning they are involved in. This involves speaking, listening and participating effectively in small and large group discussions. This helps them to learn by using new vocabulary (or words they already know more accurately) to express new ideas and new thinking. This process is difficult and requires time and support. Part of the purpose of the group work is to allow this to happen. Children will hear their peers making suggestions and having ideas about the tasks. As they join in and make their own suggestions they will work together to find a solution. This will help children succeed more independently in future tasks. The discussions with the whole class will help them to be more confident in what they are saying and thinking, and will give you opportunities to provide feedback on what you are looking for in thinking lessons. The table on page 19 sets out a progression in speaking, listening and group discussion and interaction across the primary age range.

Classroom language

Classroom language is like a dialect of English. It has particular features and implicit rules that are different from language outside of school. The way you take turns, as a pupil, is very different from the way you normally take turns in conversation, either with your friends or at home. The teacher's use of questions, in particular, is strikingly different. Questions are often heavily loaded. For example, if you ask 'Why did you write that?', a child may assume that you are challenging them because it is incorrect and that they should have put something else. In a thinking skills lesson you may be wanting them to explain the reasons for their choices, or the decisions they made about what to write down, so as to provide a model for the rest of the class. If a teacher asks 'What do you *think* you should do?', the pupils may assume that you are reprimanding them for not listening, rather than asking them to speculate. It is therefore very important to think carefully about the questions that you ask to try to ensure that your pupils understand you really *do* want to know what they are thinking! Some examples of good questions are provided on page 21.

Talking maths

Mathematical language is also different from everyday English. It is important that children do not just learn and remember the vocabulary, but learn how to use the language to communicate. This will help them to develop their mathematical thinking. Many words have specialist meanings in maths lessons, such as 'odd' and 'even'. Other words may not be encountered outside of these lessons, for example, 'trapezium' and 'numerator'. The *Thinking by Numbers* activities are a chance for children to speak the language of mathematics, rather than just practise its vocabulary.

Suggestions for further reading
Primary National Strategy, *Speaking, Listening, Learning: Working with children in Key Stages 1 and 2. Professional development materials* (Document code: DfES 0163-2004) (2004)

N. Mercer, *Words and Minds: How We Use Language To Think Together* (London, Routledge, 2000)

S. Higgins, *Parlez-vous mathematics? Enhancing Primary Mathematics Teaching and Learning*, I. Thompson (ed.) (Buckingham, Open University Press, 2003)

A skills progression in ...

	... Speaking	... Listening	... Group discussion and interaction
Y1/2	◖ Speak clearly and expressively in supportive contexts on a familiar topic. ◖ Order talk reasonably and pace well when recounting events or actions. ◖ Talk engagingly to listeners with emphasis and varied intonation. ◖ Able to use gestures and visual aids to highlight meanings.	◖ Listen actively following practical consequences, e.g.: – looking at a speaker – asking for repetition if needed. ◖ Able to clarify and retain information: – by acting on instructions – by rephrasing in collaboration with others – by asking for more specific information.	◖ Talk purposefully in pairs and small groups. ◖ Contribute ideas in plenary and whole-class discussions. ◖ Make and share predictions, take turns, contribute to review of group discussion. ◖ Review and comment on effectiveness of group discussions.
Y3/4	◖ Sustain speaking to a range of listeners, explaining reasons, or why something interests them. ◖ Organize and structure subject matter of their own choice, and pace their talk (including pauses for interaction with listeners) for emphasis and meaning. ◖ Adapt talk to the needs of the listeners (such as to visitors or more formal contexts), showing awareness of standard English.	◖ Sustain listening independently and make notes about what different speakers say, identifying the gist, key ideas and links between them. ◖ Able to comment and respond, evaluating a speaker's contribution, or evaluate quality of information provided. ◖ Able to concentrate in different contexts, including talk without/by actions and visual aids.	◖ Sustain different roles in group work (with support from a teacher), including leading and summarizing main reasons for a decision. ◖ Talk about language needed to carry out such roles and how they contribute to the overall effectiveness of the work. ◖ Reflect constructively on strengths and weaknesses of group talk.
Y5/6	◖ Develop ideas in extended turns for a range of purposes. ◖ Assimilate information from different sources and contrasting points of view, present ideas in ways appropriate to spoken language. ◖ Use features of standard English appropriately in more formal contexts. ◖ Make connections and organize thinking.	◖ Listen actively and selectively for content and tone. ◖ Able to distinguish different registers, moving between formal and informal language according to the audience, and emphasize or undercut surface meanings. ◖ Able to discern different threads in an argument or the nuances in talk.	◖ Organize and manage collaborative tasks over time and in different contexts with minimal supervision. ◖ Negotiate disagreements and possible solutions, by clarifying the extent of differences, or by putting ideas to the vote. ◖ Vary the register and precision of their language and comment on the choices made in more formal contexts.

Adapted from Primary National Strategy, *Speaking, Listening, Learning: Working with children in Key Stages 1 and 2 Handbook* (Norwich, DfES/HMSO, 2003)

Collaborative group work

Collaborative group work is an essential part of thinking skills teaching. The opportunity to work with a partner or in a small group is essential. This is where children can explore their own thinking, hear other people's ideas, be tentative, make mistakes, but be supported and encouraged by their peers. This is how an individual develops confidence in new ways of thinking. However, it does not happen automatically. You will need to make time for it, support, nurture and encourage it.

Plan for it

Thinking about who is going to work with whom, and how, is essential. It won't just happen until the class are used to this way of working, and even then there will be new skills they can develop. Most thinking skills lessons are based on mixed groups that are not based on current levels of attainment. However, you will need to monitor who works well with whom and support the children in working with a wider range of their peers.

Make it explicit

The children need to know that they are expected to work together, and that you are expecting them to help each other. This needs continual reinforcement with the whole class in the introduction, mini-plenaries and review sections of lessons (praising and reminding groups and individuals helps, too).

Teach pupils how to work in groups

Not all children find it easy to cooperate. They may well need the first few activities to focus on learning to work together. It is worth making this a part of your learning objectives for speaking and listening (see pages 18 to 19). In one of the early sessions (if you have not done so already), it is worth agreeing class rules for working in groups or a 'working together protocol'. Such an agreement should be phrased positively about what children should do and might include things like:

- Make sure everybody has a turn in speaking
- One person speaks at a time
- Look at the person who is talking (make eye contact)
- Listen actively (positive body language such as nodding or an open posture)
- Speak clearly
- Explain what you mean
- Respond to what other people say
- Make a longer contribution than just one or two words
- Give reasons for what you think
- Make it clear when you disagree that it is with what has been said (with your reasons) and not a person.

However, it is important that the precise wording comes from the children and that the agreement is posted publicly where it will always be visible in the classroom. The children will use it!

Start small

Pairs are the easiest groups to start with. In Key Stage 1 this should be the main aim. Even very young children should be able to cooperate in pairs, particularly if the cooperation is structured in some way (such as taking turns in a game). Moving from pairs to fours is a good tactic too. A paired task can be reviewed by two pairs to reach agreement, then this larger grouping can form the basis for a further activity.

Make sure the tasks require cooperation

Consider strategies such as having one recording sheet, or set of resources that need to be shared, or assign specific tasks to each member of the group. As groups get bigger you may need to assign different roles and let the children practise the different skills required (for example, leader, note taker, summarizer, clarifier). In the beginning it is best to use existing friendships as the basis for organizing the groups, but don't let them get too cosy. Learning to work with people who are not close friends is an important skill for life!

> ### Suggestions for further reading
> L. Dawes, N. Mercer and R. Wegerif, *Thinking Together: Activities for teachers and children at Key Stage 2* (Birmingham, Questions Publishing Co., 2000)

Talking points

Getting started

How are you going to tackle this?
What information have you got to help you?
What do you need to find out or do?
How are you going to do it? Why that way?
Can you think of any questions you will need to ask?
What do you think the answer or result will look like?
Can you make a prediction?

Supporting progress

Can you explain what you have done so far?
What else do you need to do?
Can you think of another way that might have worked?
What do you mean by ...?
What did you notice when ...?
Are you beginning to see a pattern or a rule?

If someone is stuck ...

Can you say what you have to do in your own words?
Can you talk me through where you are up to?
Is there something that you know already that might help you?
How could you sort things out to help you?
Would a picture help, or a table/sketch/diagram/graph?
Have you talked with your partner/another pair/group about what they are doing?

Reviewing learning

What have you learned today?
What would you do differently if you were doing this again?
When could you use this approach/idea again?
What are the key points or ideas that you need to remember?
Did it work out the way you expected?
How did you check it?

Remember – one way to ask a question is just to wait!

Suggestions for further reading
Association of Teachers of Mathematics (2004)
Primary Questions and Prompts Derby: ATM

There are a number of general teaching strategies that you can explore to support the activities in *Thinking by Numbers*. They are helpful because you can use the same technique in different contexts and develop thinking across the curriculum. Each time you use these strategies you can focus on the children's thinking that you want to develop. The children become familiar with the techniques and can get straight down to the learning involved. The strategies are also useful in assessing the children's understanding. If you first **demonstrate** a technique or approach, you can then set an activity which the children **undertake** to develop their thinking. This is as far as most approaches to thinking skills go. However, if you then set a challenge where the children have to **generate** their own activity based on what they have done, you will see them reveal their understanding of the thinking required. This cycle of **demonstrate**, **undertake** and **generate** ensures that the thinking becomes embedded.

Odd one out

In this strategy the children are presented with three items and asked to choose one as the 'odd one out' and to give a reason. Items are chosen to ensure that a range of answers is possible. Pupils can also be asked to identify the similar corresponding characteristic of the other two, or features common to all, to develop their vocabulary and understanding. In mathematics this leads naturally on to a discussion of the properties of numbers and to identifying numbers which have a range of properties. It can easily be extended to work on shapes or into other subjects. Selecting three items with different possible reasons is essential. When the children design their own game, it is essential that you emphasize that there should be more than one solution or 'answer'. It leads on to identifying common properties that the odd one out lacks.

Living graphs

The strategy involves a graph or a chart as the basis for an activity where the children have to relate short statements to the more abstract structure of a graph. The use of statements that children can understand easily, but which they then have to discuss and interpret, helps them to make sense of both the representation of the graph and the information it is based on. This works well in mathematics and science, but also in other subjects where quantitative information is used, such as history and geography.

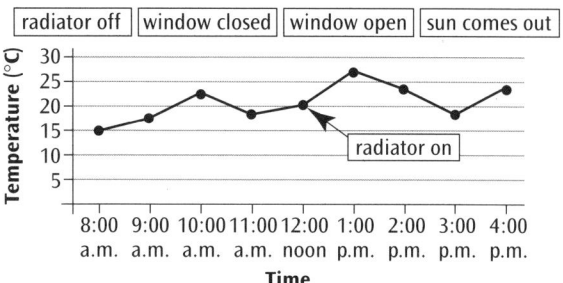

Always, sometimes, never

Another useful strategy is to have a set of statements, such as 'triangles have three sides' or 'multiples of 3 are odd' and ask the children if they are 'always' true, 'sometimes' true or 'never' true. This works well in mathematics and science: in other subjects you may need to set these categories along a continuum to provoke discussion.

As before, asking the children to make up statements that are always, sometimes or never true is a good way to extend the task (and their thinking).

Think/Pair/Share

This is a good general technique to get everyone thinking. Instead of getting a response from an individual pupil ask the whole class to work out the answer, then see if the person next to them agrees, then ask each pair to discuss what they have agreed with another pair. A further variation gets the children to record their thinking before discussing it with a partner 'Think/Ink/Pair/Share'.

Fermi questions

The approach of Enrico Fermi, who was an Italian scientist who used to pose questions to get his team thinking and working together, works well in the classroom. At school a question such as: *How many balloons would it take to fill the school hall?* requires the children to ask a number of related questions along the lies of, *How big is a balloon? How big is the hall?* This particularly develops estimation and approximation skills. Discussion and reasoning is an important part of the process of answering them. Other questions might be: *How many chocolate beans will it take to fill a litre lemonade bottle? What is the total mass of all the children in the school? Or If everyone in school (or the class) lay down in a line from the school gate (or classroom door), head to toe, where would the line end?* Once the children get used to answering questions like this you can ask them to think up their own.

Banned!

Another strategy uses an approach that involves describing an idea or object without using certain banned words (the Association of Teachers of Mathematics [ATM] have a mathematical version called 'Fourbidden'). This can be used to develop creative use of language to describe familiar ideas and concepts. This strategy works well across the curriculum and can be used to get children thinking creatively about their use of language.

describe:	without using:
a square	four sides shape equal

PMI

PMI stands for 'Plus/Minus/Interesting' and is a technique developed by Edward de Bono (as part of his Cognitive Research Trust [CoRT] programme) to get beyond the basic 'pros and cons' approach and the snap decisions that can result from this. When there is a difficult decision or where evaluation is needed, draw up a table headed up 'Plus', 'Minus', and 'Interesting'. In the column underneath the 'Plus' heading, ask the children to write down all the positive points of taking the action. Underneath the 'Minus' heading they write down all the negative points. In the 'Interesting' column they write any further thoughts that strike them. These can be scored across the class to find out how many plus and minus points there are as a method of voting.

Mind-mapping and Concept-mapping

In 'Mind mapping' the children are asked to brain storm an idea or a concept to create a web of related ideas with branches for each related sub idea. It can be used in a wide variety of ways. Mind mapping is usually done individually as a means to represent thinking on a topic, either to record ideas or so that the connections between ideas can be developed. In 'Concept-mapping' the links between the different ideas on the 'map' are labelled so that the relationship between the ideas is expressed more precisely. This supports clarification and developing understanding of the relationships between ideas. Concept Mapping has been used across the curriculum, but particularly in science, as a way of assessing change in understanding by pupils.

Activities in this book

Unit 1 Sort it out! *Information processing skills*	**Easy as pi (pages 26–29)** In this investigation children gather information and find a pattern in the relationship between the circumference and diameter of a circle. **Who wants to be a millionaire? (pages 30–33)** The children work out what they could buy with a million pounds. An example of a 'Fermi question', named after an Italian physicist who used to pose puzzles to his staff to get them to think and solve problems with limited information.
Unit 2 That's because … *Reasoning skills*	**Chains (pages 36–39)** In this activity children generate sequences of numbers by applying a simple rule. They use reasoning to work out that they have completed the task. **Braille and beyond (pages 40–43)** Children search for patterns of dots, based on the Braille alphabet. By using a systematic approach children deduce that they have indeed found every solution.
Unit 3 Detective work *Enquiry skills*	**Lost in the loop (pages 46–49)** In this enquiry the teacher proposes a plausible theory. Children conduct an investigation to test this theory and then move on to organize their findings efficiently. **Fenced in (pages 50–53)** Though children may begin this task using trial and error they will quickly find better ways to solve it. The rules of the puzzle are easy, so children will be able to see effective solutions as they tackle it.
Unit 4 What if …? *Creative thinking skills*	**Tiling to order (pages 56–59)** Children make a repeating pattern from simple tiles according to a rule that they have helped to create. When a range of ideas has been tried children can investigate the creative process itself by identifying possible modifications to the basic method. **At sixes and sevens (pages 60–63)** The children have to display given numbers on a calculator by using only a handful of keys. By extending and adapting simple starting ideas, children find that they can create new solutions that were initially far from obvious.
Unit 5 In my opinion … *Evaluation skills*	**So what? (pages 66–69)** The children identify significant facts and consider what they mean. Following discussion, children make statements that demonstrate understanding or propose an action (or both). **Domino dots (pages 70–73)** Children are challenged and guided to find a range of methods to answer the question of how many dots there are in a set of dominoes in order to evaluate the solutions.
Unit 6 Think on! *Using and applying thinking skills*	**Game on (pages 76–79)** The children design and make a game in order to apply the skills they have developed in the earlier units. **Classroom reshuffle (pages 80–83)** In this activity children combine thinking skills and practical measurement with teamwork to improve the classroom environment. **What's the issue? (pages 84–87)** This activity is about gathering data for a purpose when a real need arises to solve a particular problem.

Sort it out!

Information processing skills

> **Information processing** – these skills enable pupils to locate and collect relevant information, to sort, classify, sequence, compare, contrast and analyze part/whole relationships. (QCA 2000)

Overview

This unit is about working with mathematical ideas and concepts by gathering information. It is about building understanding by actively working with these concepts and ideas. It is about remembering links and making connections to understand what information is relevant. It is also about working with ideas to develop understanding of their meaning by working with patterns and rules, working with definitions and organizing and representing ideas. It is an essential aspect of mathematical thinking. The activities in this unit are designed to help pupils engage practically with ideas and information so as to build their knowledge and understanding of mathematical concepts.

Strategies

Information processing skills can be broken down further into the following kinds of behaviours or activities that pupils can do:

- **Find relevant information**
  Remember, recall, search, recognize, identify
- **Collect relevant information**
  Retrieve, identify, select, gather, choose
- **Sort**
  Group, include, exclude, list, make a collection or set
- **Classify**
  Sort, order, arrange *by kind or type*
- **Sequence**
  Order, arrange *by quantity/size/weight*, put in an array
- **Compare**
  Find similarities (and differences), examine, relate, liken
- **Contrast**
  Find differences/similarities, examine, distinguish
- **Analyze part/whole relationships**
  Relate, consider, sort out, make links *between parts and wholes* (e.g. component/integral object (such as the face of a cube); member/collection; portion/mass; stuff/object; place/area; feature/activity; especially in terms of fractions, ratios and the like).

Questions

Can you think of something that might help? What does this remind you of?
Give me an example of a … Is … an example? Can you give a counter-example?
What would come next? What would come before this?
Why is it the same/different? What makes it a …? What is it like? What makes a … different from a …?

Easy as pi

In this investigation the children gather information and find a pattern in the data that they collect. Groups must organize themselves well and ensure that everyone has a part to play. The children decide how much information they need to discover the anticipated pattern. Ingenuity in devising and using reliable measuring techniques will prove helpful.

Key maths links

- Measures
- Problems involving measures

Thinking skills

- Information processing skills
- Planning what to do
- Comparing

Language

length, width, perimeter, centimetre, millimetre, ruler, tape measure

Resources

PCM 1 (one per group)
PCM 2 (as needed)
three or four **large paper squares** of different sizes
circular objects of various sizes (bottles, tins and other containers)
measuring equipment: tape measures, rulers, string, calipers etc.
calculators

 Setting the scene

Display three or four large squares of various sizes cut from coloured paper (or draw them on a board or flipchart). Explain that you want to discover if the dimensions of each square have anything in common. Using a tape measure, or length of string and ruler, measure the perimeter of one square and note this on the shape, (N.B. do not calculate the perimeter by multiplying the edge length by 4). Record the width of the shape. Repeat for the other squares. *Do these measurements have anything in common?* Discuss. Draw a table to summarize the measurements:

Perimeter of square	Width of square	Linking number (perimeter ÷ width)
84 cm	21 cm	4

Calculate the value of 'perimeter ÷ width' for each shape. Call this the linking number. Of course, within the limits of your measurements the result will be 4 in each case. The children will be able to explain this result. *Would something similar happen with circles?*

 Getting started

Supply a selection of objects with a circular cross-section. Ask the children to work in small groups to undertake an investigation of the link between the 'distance around a circle' and its 'width'. Allow the children to plan and organize themselves. Let them make mistakes.

Simplify

If you judge that the children need guidance they can use PCM 1a to record results. Part of the challenge of this activity is to find ways to gather reliable information. However, if you judge children are struggling this is an ideal opportunity to teach specific techniques, e.g. using calipers or parallel straight edges for the circle's width, and wrapping string or a paper strip once or more around the object to measure its circumference (PCM 1b).

Challenge

If the children are skilled at making line graphs they can record measurements graphically (use PCM 2: the dashed lines indicate linking ratios of 3 and 4 and plotted results are likely to lie between the two). Repeat measurements and calculate average values to improve accuracy. You could use appropriate technical vocabulary: *circumference, diameter, ratio*. However do not obscure the point of the activity by introducing too much that is new.

Checkpoints

You will want to discuss the accuracy of the children's measurements. Acknowledge that 100% accuracy is impossible but emphasize the need for techniques that supply reliable results. When the children calculate the ratio linking the distance around each circle with its width they will find their calculators will display numbers with many decimal places. Consider how to handle these values with the children.

Watch out for ...

Some groups will be tempted to try to measure all the objects. If haste compromises accuracy then the children's results will be unreliable. Ensure the children work as carefully as possible. Encourage all of them to take a part in the practical work.

Ask ...

- ❍ *How many measurements will you take?*
- ❍ *What do you expect?*
- ❍ *Is there a link? Are you sure?*

Listen for ...

Applaud the children's efforts to gather accurate information: *I'm going to check that measurement.*

Moving on ...

Compare the outcome of different groups' investigations. Do the results indicate a consistent link between the perimeter of a circle and its width? What is the 'linking number' for circles? Is there much variation in its calculated value? Why? (You could explain that the linking number for circles is called 'pi' (π) and give a little more information about it. However, calculations using π are still some years away for most children and will almost certainly not be worth attempting.)
Also consider how each child contributed to the investigation. How did groups organize themselves to complete the task? Did everyone have a part to play?

Where next?

- ❍ Extend the investigation to other shapes. What is the linking number for a range of regular polygons?

Did you give the children enough freedom to direct their own investigation? Were they able to exercise ingenuity in gathering information? Did they organize their data successfully and make the looked-for link? Did groups make mistakes? And learn from them?

DEBRIEF

Easy as pi

Name _____ Date _____

Find the link between distance around a circle and its width.

Distance around circle	Width of circle	Linking number (distance around ÷ width)

✂ -

Easy as pi

Ways to measure circles.

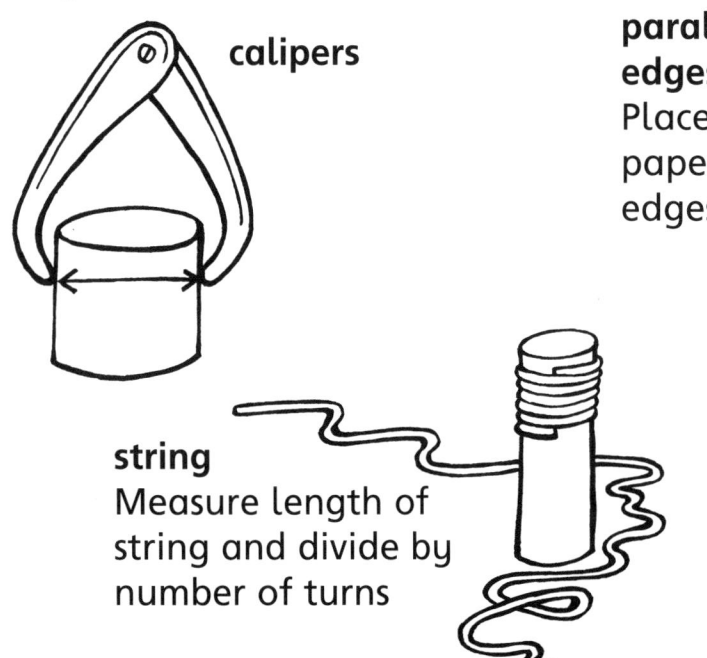

calipers

string
Measure length of
string and divide by
number of turns

parallel straight edges
Place on squared
paper to keep
edges parallel

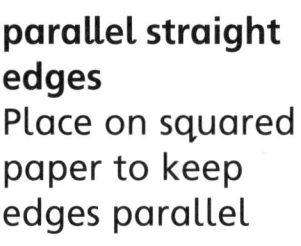

paper strip
Prick through
strip with pin

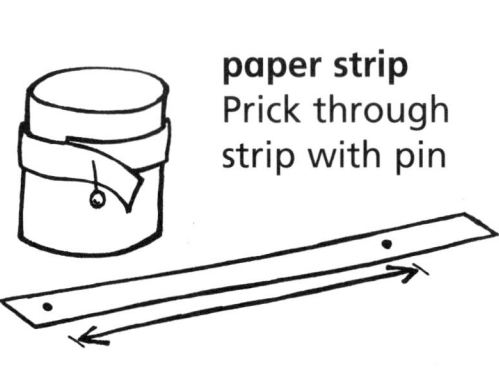

PCM 2

Finding the linking number

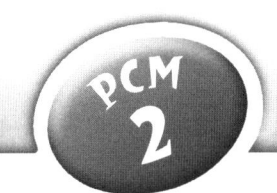

Name ______________________________ Date ________________

Record your measurements on the graph.

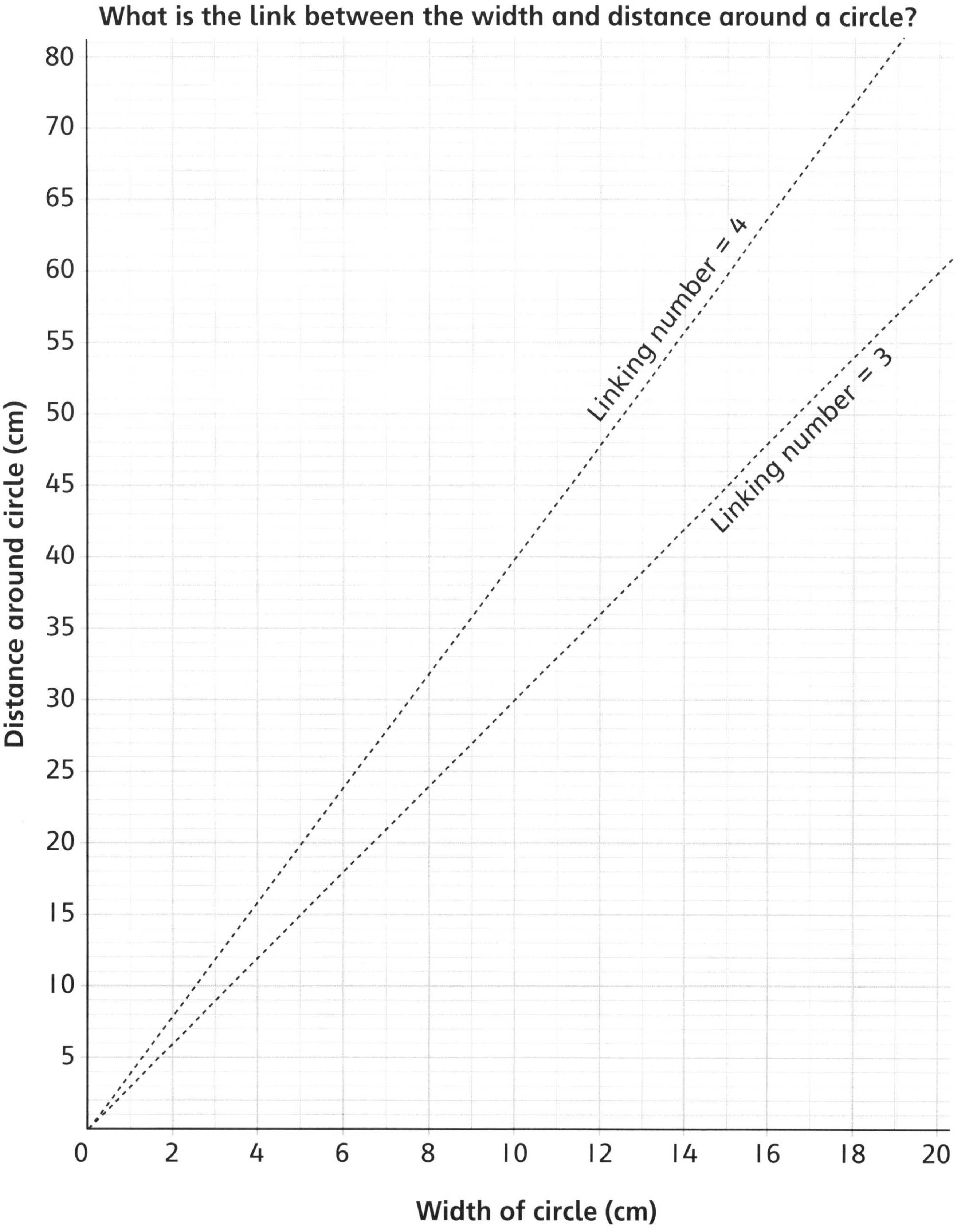
What is the link between the width and distance around a circle?
80
70
65
60
55
50
45
40
35
30
25
20
15
10
5
0
2
4
6
8
10
12
14
16
18
20
Distance around circle (cm)
Width of circle (cm)
Linking number = 4
Linking number = 3

Who wants to be a millionaire?

What could you buy with £1000 000? In this activity you pose a challenging question and encourage the children to invent sufficiently simple ways to answer it. One of your tasks is to point out the way children's strategies evolve in the light of experience. There will be setbacks and dead ends. It is reassuring for children to know that mistakes are inevitable and perfectly acceptable. Problems of this kind are sometimes called 'Fermi questions', after Enrico Fermi, an Italian physicist, who used to pose puzzles to his staff to get them to think and solve problems with limited information.

Key maths links

- Estimating and rounding
- Problems involving 'real life' and money

Thinking skills

- Information processing skills
- Planning what to do
- Collecting relevant information
- Having confidence in judgements

Language

million, approximately, sum, total, average, typical

Resources

PCM 3 (optional, as needed)
PCM 4 (one for display, or one per group)
store catalogues (preferably the same)
calculators

 ## Setting the scene

Display a mail order or High Street store catalogue. Pose the question: *If you had £1000 000 could you buy everything in here?* Discuss the children's initial thoughts, then ask: *How could you actually find out?* Ask the children to discuss, in pairs or small groups, possible ways to begin.

 ## Getting started

Before starting satisfy yourself that everyone is clear about the concept of one million. Distribute catalogues and calculators to pairs or groups of three children. Ask if you could, indeed, buy a single example of everything in the catalogue with £1 000 000. Explain that you expect a clear answer – yes or no – but you will need an explanation of how the decision has been reached for it to be believable.

Simplify

Initially many will total the cost of a page at a time with no approximation or rounding. This is hugely time-consuming. Discourage this approach and seek alternatives. The children can start with a smaller sum, say £1000 and consider progressively larger amounts. Or you can split a single catalogue and give sections to different groups to work with. Getting a feel for the cost of the goods on a typical page is another way to tackle the question. Encourage them to use PCM 3 to help (they do not need a detailed understanding of mean average to use this approach).

Challenge

The simplest way to challenge the children as they do this activity is to leave them alone. In practice, though, you will want to guide and encourage them according to their competence and confidence, promoting strategies that look promising and querying those that you suspect will prove fruitless. Wherever possible, build on the children's own ideas.

Checkpoints

Stop the class from time to time to acknowledge where existing strategies are falling short. Discuss the possible development of these strategies, or prompt the search for brand new ones. You could record the evolution of the class's attempt to solve the puzzle by using a large version of PCM 4, or alternatively make notes to share with the whole class later.

Watch out for ...

Those who persevere in slavishly totalling prices should be guided towards a better approach. Under your guidance strategies may converge on a key question, e.g. *What is the cost of the items on a typical page?* If so, engage pairs/groups that are not pursuing a promising strategy of their own in answering this question, thereby gathering the necessary information collaboratively. Use PCM 3 if it helps.

Ask ...

- ◑ *Can you think of something that might help?*
- ◑ *Do you have to work so accurately?*
- ◑ *Can you think of a quicker method?*
- ◑ *Is your answer good enough?*

Listen for ...

Children may grow dissatisfied with their early attempts: *This is going to take forever.* A fresh start may well be a good choice. Some will need reassurance that this is the best course to take.

 Moving on ...

Briefly consider the answer to the original question: *If you had £1 000 000 could you buy everything in the catalogue?* This need not take long as you will have had much discussion of this point within the lesson. Then review the process of answering the question. Consider, for example: *Was it clear what to do at first? Did you have to change your mind? Did your understanding of the puzzle improve with experience?* Chronicle the class's progress, problems, setbacks and achievements using a display copy of PCM 4, referring to your observations and notes, if you made them. With the children identify valuable lessons learnt by the class: *You weren't afraid to try a new approach when ..., It looked like a setback when ..., It was good that you were prepared to try ...,* Highlight that progress is rarely entirely smooth.

Where next?

- ◑ Estimate the actual cost of everything in the catalogue.
- ◑ Estimate how many bricks there are in the school building.
- ◑ Display a large version of your completed PCM 4 in the classroom. Refer back to its valuable lessons in other contexts.
- ◑ Use group/individual copies of PCM 4 when undertaking other tasks.

Were you able to chronicle the story of the lesson? Did it have a discernible shape? Did the children learn useful, transferable lessons? Could you use PCM 4 again to make thinking skills explicit in other contexts?

Name _____   Date _____

What could you buy with £1 000 000?

Page number	Type of goods	Cost of goods

Spending a million

Name _____ **Date** _____

How to spend £1 000 000.

Finally:

Comment:

Then:

Comment:

Next:

Comment:

To start with:

Comment:

Assessing progress

You know that children are developing their skills in information processing when they start to make connections with different mathematical ideas. They should start to show and use this understanding in other lessons. This might be by applying mathematical knowledge in a new situation or it might be in the way that they go about a subsequent task. As their skills in using information develop they should become more precise in the way that they use mathematical language and more systematic in their approach to working and to recording. The techniques that they have used should be developed in other subjects so that their understanding of information processing skills can be transferred to other areas of the curriculum.

Cross-curricular thinking

Literacy

A strategy like 'odd one out' can be used to compare characters from fictional genres, such as different heroines from traditional tales.

Art

The 'odd one out' strategy is also useful for comparing the work of famous artists or to look at similarities and differences in the visual and tactile qualities of materials.

Science

Little Red Riding Hood, Snow White and Cinderella – who is the odd one out? Why? What makes the other two the same?

Venn diagrams are powerful tools in the teaching of classification. This is particularly valuable in the strands of both variation and classification of living things, and materials and their properties.

Features of plants	similarities	Features of animals
	differences	

History

Venn diagrams are useful to teach how to compare and contrast in history. Two intersecting sets can be used as a planning tool to identify similarities and differences between different historical periods. Common features go in the intersection and contrasting information on each side.

Geography

The 'odd one out' strategy and Venn diagrams are both helpful in geography. The former can be used to encourage children to use geographical vocabulary as they talk about what makes three different landscapes or features of the environment similar or different. The latter can be used for sorting pictures of buildings or vocabulary related to the character of places. This way the children will develop an understanding of these concepts by having examples and counter-examples to talk about in a meaningful context.

That's because ...

Reasoning skills

> **Reasoning** – these skills enable pupils to give reasons for opinions and actions, to draw inferences and make deductions, to use precise language to explain what they think and to make judgements and decisions informed by reasons or evidence. (QCA 2000)

Overview

This unit is about reasoning and logical thinking. Reasoning is an essential aspect of mathematics and underpins the development of theorems and proofs through the use of precise definitions and axioms. For pupils of primary age it is important that they have the opportunity to apply their knowledge and understanding of mathematical ideas and concepts logically and systematically as this will enable them to make connections between different concepts and between different areas of mathematics. This will deepen their understanding and develop their confidence as well as helping them see how mathematics can be used as a practical tool in their daily lives.

Developing reasoning skills is also about developing habits of thinking or dispositions as much as it is about specific logical skills. Of course, just because you are good at reasoning does not mean that you are going to be reasonable. Part of thinking reasonably is also dependent upon your knowledge of yourself and the situation in which you find yourself. This metacognitive dimension is essential if you are going to help your pupils become effective thinkers and not just logical.

Strategies

Reasoning skills can be broken down further into the following kinds of behaviours or activities that pupils can do:
- **Give reasons for opinions and actions**
  explain, say because, say why
- **Draw inferences and make deductions**
  see links, make connections, infer, deduce, use words like 'so', 'then', 'must be', 'has to be'
- **Use precise language to explain what they think**
  exemplify, describe, define, characterize
- **Make judgements and decisions informed by reasons or evidence**
  form an opinion, determine, conclude, summarize, *especially where there is more than one course of action or possible solution*

Questions

Explain why ...? Can you give a reason ...? Because ... So ...
Why is ... an example? Is that always/sometimes/never true? What else must be true if ... ? Does it have to be like that? Can you define that? What do they all have in common?
What else is like that? What makes you say that? How can you be sure that ...?

Chains

BRIEF

In this activity the children generate sequences of numbers by applying a simple rule. There is just one complete solution. If, after attempting the puzzle, you ask the children, *Have you finished?* you will receive a range of answers, from an honest *Don't know*, to a tentative *Yes*. The challenge for you is to draw the children to understand that by reasoning they can *know* that the activity is complete. You will have succeeded when the children can say with confidence, *There are no more solutions*, and can tell you why.

Key maths links

- Properties of number sequences
- Rapid recall of addition facts

Thinking skills

- Reasoning skills
- Ordering information
- Explaining
- Giving reasons for conclusions

Language

addition, sum, total, tens boundary, remainder, proof, because, so

Resources

PCM 5 (optional, one per child/pair)
PCM 6 (for display or one per pair)

1 Setting the scene

Explain you are going to make a chain of numbers, linked together by a simple rule. Write a pair of numbers on the board or flipchart, e.g. 1 and 5. Draw a link round these numbers. *This is the first link in the chain.* Continue: . By discussion and prediction establish the rule (to calculate each term add the preceding two terms, recording only the units value if the total exceeds 10). State the rule in such a way that every child can understand and use it.

2 Getting started

Ask the children to continue the chain already begun or begin a new chain with another starting link (i.e. one pair of single-digit numbers). They can use PCM 5. If more than one chain is recorded on this sheet mark the start of each new chain clearly. After a while they will notice that sequences of numbers repeat. Gather the class together. Demonstrate the repeating behaviour of the chains. *When the pattern repeats is there any point in continuing to add links?* After a discussion conclude that there is not. Encourage the children to give reasons. *Is the investigation complete?* Through discussion establish that unfound chains probably remain. Challenge the children to find them.

Simplify

Check for the accurate application of the rule. Ask the children to work in pairs to help check calculations. Encourage clear statements of why calculations should stop when sequences repeat.

Challenge

Encourage an analytical approach. Guide the children to develop methods that help them identify untried starting links or that eliminate possibilities in a systematic way. Challenge them to consider if they know if they have finished or not.

3 Checkpoints

After some time stop the class and compare solutions. *Can we know when we have finished completely?* Display or distribute PCM 6, which shows the whole of the longest chain plus an ordered record of all possible links. Together, check off some links from the longest chain against the record (three have already been done).

Take care not to miss the links that span two rows (e.g. ⟨8 1⟩). Next, ask the children to check off links from their own chains. Do they have any unused links? If so use these to start new chains. When all links have been tried the investigation *must* be complete. Establish this fact in discussion with the class.

Watch out for ...

Check that any calculations are completed accurately: valid reasoning will be impossible if solutions are unreliable.

Ask ...

- *Is it worth continuing? Why? Why not?*
- *Have you finished? Are you certain? Why? Why not?*

Listen for ...

Be alert to signs of systematic work: fruitful organizing strategies can be shared and developed by others. Applaud and highlight the children's attempts to draw conclusions or offer reasons at all stages of the investigation.

 Moving on ...

The complete solution comprises just 6 chains. They are: 1 → 5... (60 links); 2 → 4 (20 links); 0 → 5 (3 links); 1 → 3 (12 links); 2 → 6 (4 links); 0 → 0 (1 link) – of course you can begin anywhere within each chain: *possible* starting links are given here. Ask: *Is there a connection between the total number of links in all 6 chains and the number in the 'possible links' list?* Both equal 100. Help the children appreciate that this means they can conclude that the investigation is undoubtedly complete. They could write up the investigation, recording solutions and explaining reasoning (*I knew when to stop because ... I knew which starting links to try by ... The investigation was over when ...*).

Where next?

- Instead of adding adjacent numbers, find their difference (this does not create simple chains - careful recording will be necessary).
- Many simple geometrical puzzles allow the children to reason to a complete solution (see 'Braille and beyond' (Unit 2, Activity 2)).
- The use of a grid or matrix for a systematic approach can be applied more widely.

Did you succeed in helping the children understand that the investigation was complete? Were you satisfied that the children appreciated this for themselves or did you feel that you had simply to tell them? You probably intervened repeatedly, working with the whole class to develop systematic working and valid reasoning. Did you time these interventions well? Did you notice the children making sense of your chain of reasoning? Did the children offer alternative reasons of their own?

DEBRIEF

Chains

Name _____ Date _____

Make your own number chains.

Links in a chain

Name _____ Date _____

This shows the links in this one long chain.

1 5 6 1 7 8 5 3 8

1 9 0 9 9 8 7 5 2

7 9 6 5 1 6 7 3 0

3 3 6 9 5 4 9 3 2

5 7 2 9 1 0 1 1 2

3 5 8 3 1 4 5 9 4

3 7 0 7 7 4 1 5 6

Fill in all the possible links. Tick off all the links that you have found in the chain.

0 0	1 0	2 0	3 0	4 0	5 0	6 0	7 0	8 0	9 1
0 1	1 1	2 1	3 1	4 1	5 1	6 1 ✓	7 1	8 1	9 1
0 2	1 2	2 2	3 2	4 2	5 2	6 2	7 2	8 2	9 2
0 3	1 3	2 3	3 3	4 3	5 3	6 3	7 3	8 3	9 3
0 4	1 4	2 4	3 4	4 4	5 4	6 4	7 4	8 4	9 4
0 5	1 5 ✓	2 5	3 5	4 5	5 5	6 5	7 5	8 5	9 5
0 6	1 6	2 6	3 6	4 6	5 6 ✓	6 6	7 6	8 6	9 6
0 7	1 7	2 7	3 7	4 7	5 7	6 7	7 7	8 7	9 7
0 8	1 8	2 8	3 8	4 8	5 8	6 8	7 8	8 8	9 8
0 9	1 9	2 9	3 9	4 9	5 9	6 9	7 9	8 9	9 9

Braille and beyond

BRIEF

In 'Braille and beyond' the children search for patterns of dots, based on Louis Braille's alphabet for the blind. By using a systematic approach they deduce that they have indeed found every solution. Whilst solving the puzzle children are encouraged to attend to the way they work and the things they say. When the children talk about what they are doing they notice the processes involved and build up a repertoire of thinking skills from which they can select in the future.

Key maths links

- Shape and space
- Reasoning about shapes

Thinking skills

- Reasoning skills
- Working systematically

Language

pattern, triangle, isosceles, scalene, square, rectangle, symmetrical, translation, rotation, reflection, because, so

Resources

PCM 7 (one per group, more if needed)

PCM 8 (one per group, more if needed)

 Setting the scene

Tell the children about the Braille alphabet. (You could retell the story of its invention, show a sample of Braille writing or write a message in Braille ...) Explain that each letter of the alphabet is represented by a pattern of raised dots on a 3×2 rectangular grid. (Use the Internet to find the complete Braille alphabet). Indicate that you want them to discover the different patterns that can be made using the Braille system. In addition, state that you want them to record how they attempt the challenge. Explain that group members will take it in turns to make notes. Model the use of PCM 8 to capture observations. Encourage the children to speak their thoughts aloud. The person acting as scribe may need to keep asking: *What are you doing?*

 Getting started

Organize the children in groups of three. Two of the children use PCM 7 to record solutions whilst the third uses a copy of PCM 8 to record observations. At your signal the children swap roles. (You can entrust this responsibility entirely to the groups but there is a danger that it will be overlooked. Your active oversight will improve the chance that the process of solving the problem is made available for discussion.) Encourage the children to discover systematic ways of working.

Simplify

Act as scribe to capture the children's developing strategies (instead of, or as well as, the children themselves). If it helps, suggest the children initially use a 2×2 grid to develop a systematic approach; then extend to the 3×2 Braille arrangement.

Challenge

Guide the children to find more than one strategy and then compare them. Challenge them to develop a precise vocabulary to describe what they are doing: *I'm arranging all the scalene triangles first; I've found all the 2 dots, now the 3 dots ...*

 Checkpoints

Some children may start without a system. As you challenge them to find all possible solutions they will discover the need for one. They could: place dots on the grid sequentially; focus on the shapes made by the dots (e.g. isosceles triangles), translating, reflecting and rotating these to find all orientations; label the grid and write an ordered set of numbers to describe the solutions; use arguments from symmetry to simplify the task.

Watch out for ...

Help the children to check their solutions: it is difficult to identify incomplete solutions, and spotting duplicates is not straightforward. If children cut out their individual solutions they can regroup them systematically but they must not forget which way up each one belongs.

Ask ...

- ◗ *What is your method?*
- ◗ *Are there other solutions like this one?*
- ◗ *How can you be sure that you've found all of this type?*

Listen for ...

Find time to make your own observations. This will inform your discussions with the groups and the whole class. Listen out for examples of children's invented vocabulary: *I'm searching for all the corner-ones; The family of 1s and the family of 5s are basically the same.* Comments like these give particular insight into the organization of children's thoughts and their own lines of reasoning.

 Moving on ...

Share children's observations of how they tackled the puzzle. Add your own observations. Highlight the importance of *reasoning* (to organize and simplify the task) and of *precise vocabulary* (to communicate effectively). Provide the opportunity for the children to produce a best record of their solutions. Mount a wall display combining these along with examples of  some of the children's draft solutions plus large versions of salient comments and thoughts captured on copies of PCM 8. This display makes visible the processes involved in solving the problem. It can be the focus of subsequent discussions, perhaps preceding another lesson where thinking skills are explicitly required.

Where next?

- ◗ Apply previously gained understanding by repeating the investigation with a 3 × 3, or a 4 × 2, grid.
- ◗ Use PCM 8 in other contexts.
- ◗ Assign specific roles to group members in other investigations to make thinking strategies explicit.

Did you manage to make the children aware of how they attempted the investigation, opening their eyes to the need for systematic work? Did you help them appreciate the need for precise vocabulary, perhaps by giving status to their own informal descriptive language? How can you reinforce the thinking skills that you identified with your class? Do you think the children will be able to draw on these skills again?

DEBRIEF

Dotty patterns

Name _____ Date _____

How many different patterns can you make?

Name _____ Date _____

What we did and what we thought while making our patterns.

Assessing progress

You know that children are developing their reasoning skills when they start using words like 'because', 'then' and 'so' in their discussions and their responses to your questions. They may also start to ask each other 'why?' questions and seek explanations from each other (and from you). Giving reasons as part of explanations then becomes a routine part of thinking lessons. Once you start to ask children why (or to ask another child why a response either was or was not correct) you will be able to assess the reasons in their responses. You need to ensure you ask children to justify correct and incorrect responses otherwise they will 'read' your question as meaning they have made a mistake if you only ask 'why?' when an answer is wrong. Once children get used to this you can simply wait encouragingly or say 'because ...?' to get them to extend their replies to your questions to assess their reasoning skills.

Cross-curricular thinking

Science

Asking a question such as, *What will happen if?* is a good starting point for scientific reasoning. *What will happen if you put a tea cosy over an icy drink? Will it warm up faster or more slowly? What will happen if you drop a football and a cannonball from the top of a tall building? Will the cannonball reach the ground first?* Using thought provoking questions like these can stimulate scientific reasoning (as well as revealing children's thinking about scientific concepts).

History

A strategy like 'odd one out' can also be used to develop reasoning skills as the children are asked to give reasons for their choice of an 'odd one out' and can be encouraged to distinguish between historical and non-historical reasons. Choose three famous people and ask children to identify an odd one out with a historical reason.

Literacy

Justifying choices of words and phrases is a good way both to develop reasoning, and model thinking about composition. Asking a series of questions such as, *Why did you choose that adjective or powerful verb? What others did you consider? Why did you reject those?*, not only gives children the opportunity to give their reasons, but to make them explicit for others to hear.

Geography

Geographical enquiry is supported by reasoning as children express their views about places or changes to the environment. They can use a technique, such as identifying 'Plus, minus and interesting' points to compile a table, then justify the points they have identified with reasons.

Detective work

Enquiry skills

> **Enquiry** – these skills enable pupils to ask relevant questions, to pose and define problems, to plan what to do and how to carry out research, to predict outcomes and anticipate responses, to test conclusions and improve ideas. (QCA 2000)

Overview

Enquiry skills are as much a way of working or developing particular habits of mind which keep a range of possibilities open for as long as possible. The process of enquiry is about being flexible, looking for alternatives and testing a range of possible solutions. In mathematics these are essential skills as enquiry develops an understanding of relationships and connections that may not be immediately obvious.

The process of enquiry is at the heart of learning. It is only when you can identify what you need to know, go through a process of finding out and be able to recognize when you have found a solution that you can undertake independent learning. Enquiry skills can, therefore, best be developed in situations where it is not possible to see a solution from the outset and where children will benefit from working together.

There are good opportunities for speaking and listening in presenting the results of an investigation or enquiry. Enquiry lessons are also excellent for review and reflection about the process of learning.

The challenge for the teacher is at the beginning and end of the enquiry process. It is difficult to instruct children in how to ask relevant questions without directing them to a particular investigation or mathematical problem. Similarly, it is difficult enough for pupils to recognize that they have come up with a solution to an investigation, without them realizing that it is a good solution. Identifying what would be a better answer is even more difficult - challenging even for adults! Enquiry skills are also, therefore, about developing more systematic habits of questioning as well as the specific skills in solving a problem.

Strategies

Enquiry skills can be broken down further into the following kinds of behaviours or activities that pupils can do:
- **Ask relevant questions**
  Enquire, be curious, ask, probe, investigate
- **Pose and define problems**
  Frame, propose, suggest, put forward an idea
- **Plan what to do and how to research**
  Think out, plan, sketch, formulate or organize ideas
- **Predict outcomes and anticipate responses**
  Suppose, predict, guess, estimate, approximate, foresee
- **Test conclusions and improve ideas**
  Experiment, test, improve, refine, revise, amend, perfect

Questions

Show me how you could ...? What might work? What ideas have you got? What is a good question to ask? How could you find out? How could you check? Any predictions? What is your best guess? What are you expecting? About how much will it be?

Lost in the loop

BRIEF

In 'Lost in the loop' the teacher proposes a plausible theory. Children conduct an enquiry to test this theory and then move on to organize their findings efficiently.

Key maths links

- Mental calculation strategies
- Rapid recall of multiplication facts

Thinking skills

- Enquiry skills
- Ordering information
- Giving reasons for conclusion

Language

square number, sum, theory, prove, disprove, counter-example

Resources

PCM 9 (one per child/pair)
PCM 10 (optional, as needed)
large sheets of paper (A3 or larger)
calculators
table of square numbers

1 Setting the scene

Write the following numbers on a board or flipchart: [1 1]→[2]→[4]→[1 6]→[3 7]→[5 8]. Explain how the sequence is made: *To make the next term, add together the square of each digit in the preceding term.* Continue the number sequence: [5 8]→[8 9]→[1 4 5]→[4 2]→[2 0]→[4]... until the children notice that a pattern is being repeated. State that you have a theory that all starting numbers eventually lead to this looping pattern. Explain that you want children to confirm or deny this theory.

2 Getting started

Ask the children to begin an enquiry to test the truth of your theory that all numbers eventually get 'lost in the loop'. To build up your case direct the children to certain starting numbers, e.g. [5 7], [3 5], [2 2], [7 8], [3]. Use PCM 9 to reinforce the rule that generates the number sequences. Alter this record slightly to accommodate the occasional 3-digit number and replace it with an abbreviated style once the process is well understood. Encourage discussion and comparison of results. Ask: *Because of what you have found do you believe that all numbers get lost in the loop?* Some children may feel that this looks likely but others will express doubts: *We haven't tried all possible numbers yet.* Agree that the search is worth continuing. In time children will discover starting numbers (e.g. [7 9], [7], [2 8], [4]) that show a different outcome: such numbers lead to 1. Concede that your theory is false. Acknowledge that just one counter-example is enough to disprove it.

57   49 + 16 = **65**
25 + 49 = **74**   36 + 25 = ...

Simplify

Supply a table of square numbers to speed up calculations.

Challenge

Encourage more sophisticated recording strategies. Find as many numbers as possible that do not get lost in the loop.

3 Checkpoints

Once the falsity of your theory is accepted the enquiry changes its character. From this point on there are two distinct challenges. The first is to organize findings in a way that is at once clear and comprehensive. The children can develop their own means of summarizing their findings (large sheets of paper are useful and allow the children to display their solutions for discussion, or PCM 10 will offer guidance if needed). Secondly, the children can actively seek exceptions to your rule. By applying reasoning skills some will be able to work backwards to identify numbers that do *not* get lost in the loop. (To make the task manageable you might want to limit the search to all such numbers below 50, or 100, say.)

Watch out for ...

Check that calculations are completed accurately. If the children work in pairs then errors will be identified more efficiently. More importantly this will encourage discussion and prediction.

Ask ...

- *What are you expecting to happen? Why?*
- *How can you present your findings clearly?*
- *Can you predict if a number will lead to the loop or not?*

Listen for ...

Good teamwork will be shown by pairs/groups that divide the labour effectively and systematically: *You do one to five, I'll do six to ten.* It may avoid confusion if just one team member assumes responsibility for adding findings to the group's cumulative record. Time will be saved by children who recognize that there are usually two numbers that produce the same outcome: *I don't need to do forty-three: I know it leads to twenty-five because I've already done thirty-four.*

 Moving on ...

Review the early part of the lesson. Did the children feel the theory you offered was confirmed to start with? Ask: *How many results would you need to prove the theory? An infinite number?* Then consider the impact of finding the first exception: *How many results were needed to disprove the theory?* Note the overwhelming significance of just one counter-example. To conclude, invite groups of children to show how they have presented their findings (having used poster-sized sheets of paper will help). Select examples that are clear and complete and ask the class to comment.

Where next?

- How many steps away from the loop or finishing point is each starting number? How can this information be presented clearly? Is there any pattern?
- Conduct similar enquiries into other rule-driven number sequences, e.g. 'If the preceding number is even find the next by halving it; if it is odd then multiply it by 3 and add 1'.

Although many children have an intuitive sense of what a proof is, the concept is not trivial. However, disproving something is often clear-cut. Did you manage to convey the devastating power of just one counter-example to your cherished theory? Finally, did the children find recording their work clearly made them feel that they had mastered this enquiry?

DEBRIEF

Lost in the loop

Name _____ Date _____

Use this template to make your own number sequences.

Start here

Links in the loop

Name _____   **Date** _____

Show the links you found.

3
9
81
65
61
37
16   58
2   4   89   85 ← 29 ← 25 ← 5
11   20   145
42

Find numbers that lead to I.

1 ← 10 ← 13

Fenced in

Although the children may begin 'Fenced in' by using trial and error they will quickly find better ways to solve it. The rules of the puzzle are easy, so the children will be free to notice, and comment on, how they tackle it. As teacher your role is to listen to the children's descriptions of what they are doing, teasing out with them what theories they have invented to help them come up with the best possible answer.

BRIEF

Key maths links

- Measures
- Shape and space

Thinking skills

- Enquiry skills
- Suggesting hypotheses
- Experimenting
- Refining

Language

width, breadth, area, covers, cm², edge, symmetrical

Resources

PCM 11 (one per pair, on card if possible)
PCM 12 (optional, one per pair)
cm-square grid paper
Cuisenaire rods
(alternative to PCM 11)
scissors

1 Setting the scene

Display the square grid and fences used in the activity. An effective way to do this is to project an acetate copy of a sheet of cm-square paper on an overhead projector and place Cuisenaire rods on it. Alternatively, pin cardboard cut-outs of PCM 11 on a board. Explain that the challenge is to fence in an area. Fences with a total length of up to 48 cm can be used. Place each fence so that it is lined up with an underlying square grid. The following joins are allowed:

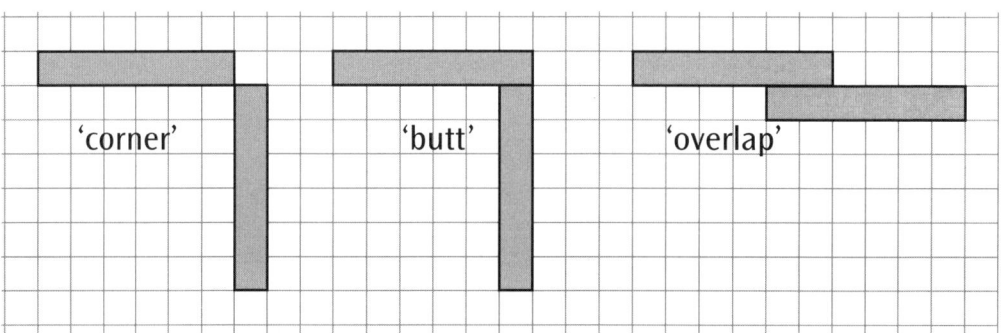

'corner'   'butt'   'overlap'

What is the maximum area that can be fully enclosed?

2 Getting started

Working in pairs, the children prepare the necessary resources. If you have Cuisenaire rods or the like, use these. Otherwise, the children can cut out copies of PCM 11, colouring them in beforehand if desired (it will help identify the different fences if they are colour-coded). Ask the children to experiment with fences in different arrangements in an effort to maximise the enclosed area. Remembering which is the best arrangement is important; comments like, *I managed 180 cm² but I can't remember how*, only lead to frustration. Use sketches or sheets of cm-square grid paper or copies of PCM 12 (for scaled-down drawings) to record the best solutions.

Simplify

Disallowing 'corner joints' makes the activity markedly easier, but less interesting.

Challenge

You can stipulate additional rules if you like, e.g. *You must use a single type of fence for each solution* or *You must use three or more types of fence*.

Checkpoints

Watch out for ...

Counting the number of square centimetres enclosed by the fences can be laborious. Encourage short cuts: block the areas into large rectangles and provide calculators to speed things up.

Ask ...

- ● *What shapes do you think make the biggest area? Why?*
- ● *How do you know where to place the fences?*
- ● *Is one type of join better than another? Why?*

Listen for ...

Some children may find it hard to describe the ideas that guide them in this task. However, the comments they make to each other will betray self-made 'rules' that direct their choices. Listen out for things like: *Don't use overlap joints, they waste fences; Long shapes are useless, they're too thin; Squares are good*, etc. Enter into conversations to tease out the theories that are at work.

Moving on ...

Consider together the strategies used in this activity. *Did you have an idea or theory that helped you choose what to do?* Listen to the children's comments. Ensure that they understand their classmates' comments: they may recognize their own ideas in others' explanations, or see something brand new that they had not even considered.

Where next?

- ● Many puzzles require children to apply a guiding principle or theory. Make this explicit by asking children to put it into words.
- ● Investigate how to play noughts and crosses so that you never lose.
- ● Use sheets of stiff paper or card (e.g. A4 size) to construct cuboids. How large a capacity is possible?

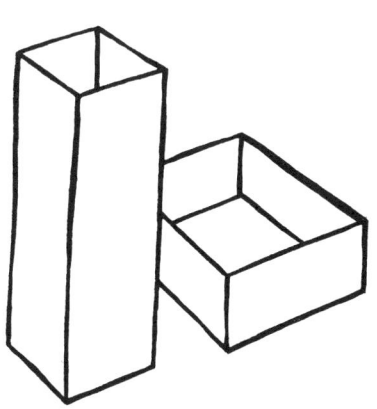

Were the children able to describe what they were doing? Were you able to turn their descriptions into explanations? Did the children understand that their choices were guided by intuitive theories about what would work well? Is it helpful to children to uncover what lies behind their thinking? Or is it better just to let them get on with it?

DEBRIEF

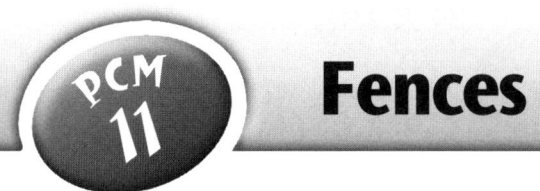

PCM 11 Fences

Colour and cut out these fences. Use them to enclose your area.

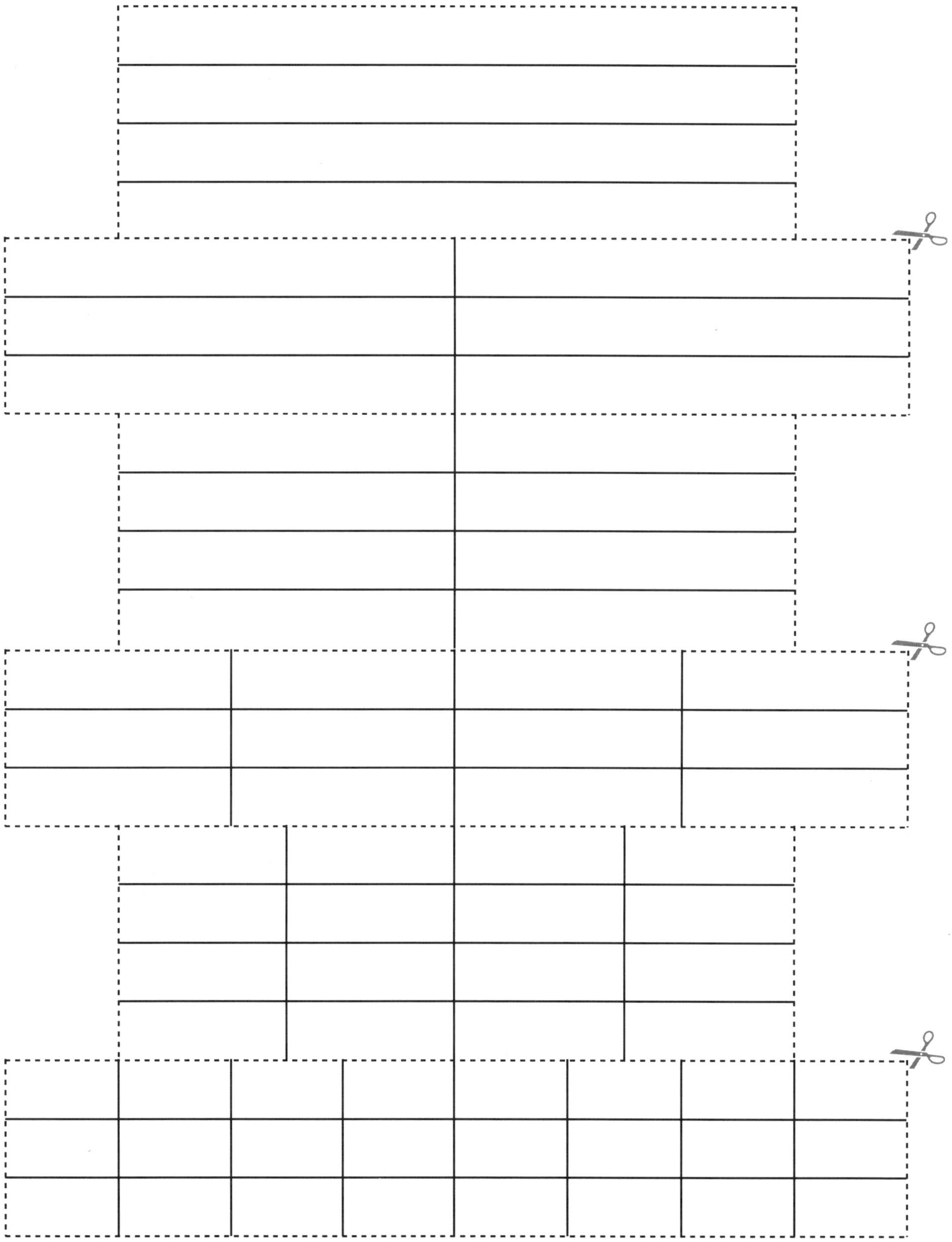

Name _____ **Date** _____

Record your areas using these grids.

Summary

Assessing progress

Evidence that children are making progress in developing enquiry skills can be gained by observing the way that they are working. Enquiry is as much a habit, or an attitude, of keeping a range of possibilities open for as long as possible. Being flexible, looking for alternatives and testing a range of possible solutions are therefore good indications that enquiry skills are developing.

Cross-curricular thinking

Literacy

Another variation on the 'Living graphs' (page 22) strategy is developing the understanding of narratives, both in fiction and non-fiction texts (such as historical narratives), through discussion and enquiry. The graph is replaced with a 'fortune line' about a character's feelings or mood. The children place statements from the narrative on the graph. To do this they need to sequence the text and empathize with the character. Investigating a number of similar narratives (such as traditional tales) will show that they tend to have a similar shaped graph, reflecting the narrative structure and the use of repetition to develop suspense (in *The Three Billy Goats Gruff* and *Little Red Riding Hood*, for example).

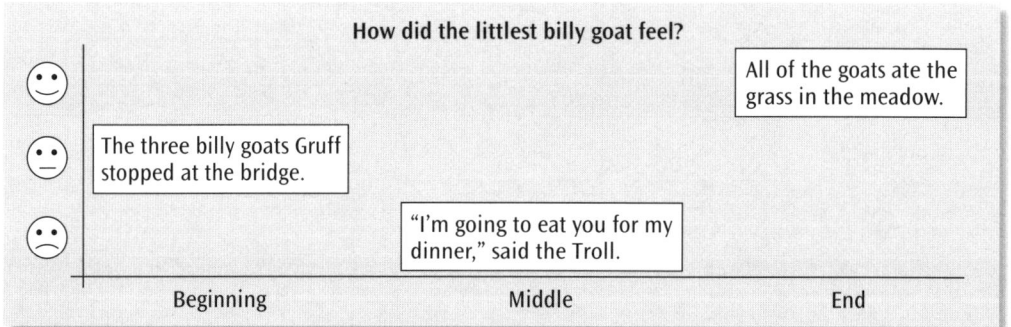

How did the littlest billy goat feel?

All of the goats ate the grass in the meadow.

The three billy goats Gruff stopped at the bridge.

"I'm going to eat you for my dinner," said the Troll.

Beginning   Middle   End

History

Fortune lines can also be used in historical enquiry particularly to develop empathy, the statements can either come from real historical figures (the diaries of Samuel Pepys and Anne Frank are good sources) or characters created for the task (such as a child miner in Victorian times).

Science

Developing scientific enquiry means the children must think up questions that can be investigated. An approach called 'Philosophy for children' has been shown to encourage children to develop questioning skills. It uses a stimulus as a starting point, commonly a familiar story, but it can be a poem or a picture, that the children think up questions about. They then select one to answer in a class discussion called a 'community of enquiry'. It is possible to extend this into science where questions can be investigated and you can challenge the children to work out how they could find out the answer.

Geography

This approach can work in geography, particularly when a photograph of an interesting landscape is used as the starting point. After a discussion of what the children think, their motivation to find out is likely to be enhanced.

What if ...?
Creative thinking skills

> **Creative thinking** – these skills enable pupils to generate and extend ideas, to suggest hypotheses, to apply imagination and to look for alternative, innovative outcomes. (QCA 2000)

Overview

Creative thinking is the kind of thinking that produces new insights, approaches, or perspectives. It is essential in education that learners see that they can come up with new ideas or suggestions which help their own thinking as well as stimulating the thinking of others. No one expects a 7- or 11- year-old to come up with something unique in the history of human development, but unless we value the creativity that young children naturally have they will stop thinking creatively and rely on reproducing ideas they have been given by others.

Creativity is often *not* associated with mathematics in schools, but thinking up new solutions to problems, seeing new connections, or thinking of more efficient or effective alternatives is what mathematicians do. It is not necessary for the ideas to be completely original, just new for the individual pupil or shared with the class for the first time, or it might be that ideas or concepts are seen in a new or unusual way. It is important that pupils feel comfortable in order to be creative. They need to have confidence that their ideas will be accepted and that there is a range of possible answers or solutions to a problem or issue. The aim is to encourage pupils to think up a range of ideas, to have new thoughts or ideas (at least for them) or to extend and develop other people's ideas.

There are a number of techniques and approaches to support creative thinking such as brainstorming, thinking of analogies, visualizing or picturing possibilities. What all these techniques have in common is an emphasis on the flow of ideas. This means that in the early stages of supporting creative thinking it is essential to be uncritical to ensure that thinking is not too restricted.

Strategies

Creative thinking skills can be broken down further into the following kinds of behaviours or activities that pupils can do:
- **Generate and extend ideas**
  Brainstorm, think up, develop, extend
- **Suggest hypotheses**
  Suppose, surmise (*use phrases like 'how about ...?', 'it could be ...'*)
- **Apply imagination**
  Design, devise, visualize, elaborate
- **Look for alternative, innovative outcomes**
  Think laterally, fancy, guestimate, invent

Questions

Can you imagine? What would that look like? How could you change it to make it a ... ? Can you think of a question you could ask? Go on ... What will the answer look like? Another idea? And another ...

Tiling to order

BRIEF

'Tiling to order' offers many creative possibilities within a supportive structure. The children make a repeating pattern from simple tiles according to a rule that they have helped to create (you can give them as much or as little freedom in this as you wish). When a range of ideas has been tried the children can investigate the creative process itself by identifying possible modifications to the basic method. In doing this they extend a simple starting idea in a number of innovative ways.

Key maths links

- Shape and space

Thinking skills

- Creative thinking skills
- Applying imagination
- Generating ideas
- Extending ideas

Language

pattern, regular, irregular, parallel, perpendicular, rotate, rotation, translation, reflective symmetry, axis of symmetry, suppose

Resources

PCM 13 (one per pair)
PCM 14 (one per pair)
lightweight card
scissors

1 Setting the scene

Draw a blank 4 by 4 addition grid on a board or flipchart [see 1 below]. Fill in the grid in modulo 4 (add a pair of numbers but record only the remainder after dividing the total by 4) [2]. Establish a code linking the digits 0 to 3 with four orientations of your chosen base tile [3]. Make a 4 by 4 'super-tile' based on your addition square, matching tiles to numbers according to your code [4]. Then arrange previously prepared copies of this super-tile side by side to make a larger design. At each stage point out how new, unexpected shapes are created where black and white areas butt against each other.

1

+	0	1	2	3
0				
1				
2				
3				

2

+	0	1	2	3
0	0	1	2	3
1	1	2	3	0
2	2	3	0	1
3	3	0	1	2

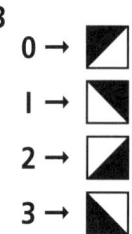

3

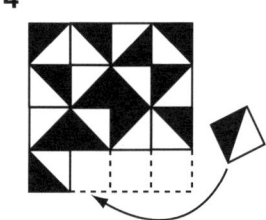

4

2 Getting started

Ask the children to talk with a partner to identify one thing they could alter in the method already described when creating a pattern of their own. Review some of the possibilities. Provide photocopied tiles (PCM 13) for children to cut out and stick in place on card. Allow plenty of time for the children to create their own super-tiles and then larger patterns. (The most efficient way to make larger patterns is to arrange to make photocopies of the children's super-tiles, copied *during* the lesson if possible to keep interest high.)

Simplify

Build confidence by allowing the children to *copy* the stages you demonstrated before they add a variation of their own. If the modulo 4 addition grid is off-putting ask the children to *design* their super-tile (i.e. don't use a rule for this part of the activity). Make the super-tile smaller (e.g. 3 by 3 grid or 2 by 2 grid). Combine the super-tiles with very simple rules (e.g. translation alone).

Challenge

Give the children free rein to alter whatever they wish, e.g. the shape of the base tile (any tessellating shape is possible); the pattern on the tile; the operation used to create the modulo grid (e.g. multiplication); the way in which super-tiles are combined to make the larger designs (these too can be arranged according to a rule rather than simple repetition).

Checkpoints

Ensure that the children consider a range of possible modifications to the procedure that you demonstrate. Encourage discussion, but allow individuals to make their own patterns if they wish (this will *generate a wide range of outcomes*). Check that the children are clear about the rules that they have established. Check that they stick to them closely from start to finish. Show good examples of super-tiles and larger patterns as they emerge. Ask the children to describe and comment on their own and others' designs.

Watch out for ...

Check that tiles are arranged neatly – careless cutting out and sloppy organization will spoil the patterns (use alternatives to glue and the results will be tidier).

Ask ...

- ❍ *What could you change when you make your pattern?*
- ❍ *Is it best to change just one thing? Or many things?*
- ❍ *What are your rules?*

Listen for ...

Notice and applaud creative thinking: *We could change this as well.* Celebrate children's precise use of the vocabulary of shape and position: *Look, a parallelogram. My super-tile has an axis of symmetry here.* Enjoy the element of surprise as patterns emerge: *I didn't expect that to happen!*

Moving on ...

Review *what* can be changed, and *how*. In the light of experience ask small groups of children to pool ideas and then generate new ideas by *extending* their thinking in a systematic way. The children could record and extend their thoughts on PCM 14 or *you* could record discussions on a board or flipchart.

Where next?

- ❍ Repeat the activity (on another occasion), using some of the novel ideas generated in the plenary discussion.
- ❍ Think of ways to introduce colour (if children don't suggest this themselves).
- ❍ Make a printing block from expanded polystyrene sheet and print a repeating pattern.
- ❍ Print patterns for a purpose (e.g. gift-wrap).

Were you able to lead children to be analytical about the possible modifications they could make to the method demonstrated at the beginning of the lesson? Were they able to extend the starting idea, or even generate innovative ideas of their own without losing the essential structure of the activity? Do you think this activity merits repetition?

DEBRIEF

Cut out and make your own super-tiles and tile designs.

Thinking by Numbers 5 • **Unit 4: What if …?** • **Creative thinking skills**

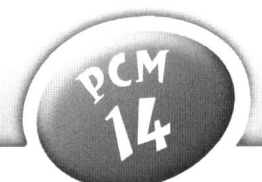

My ideas map

PCM 14

Name ______________________________ Date ________________

Record your thoughts on this ideas map to show the things you could change. Extend the map to think up new ideas.

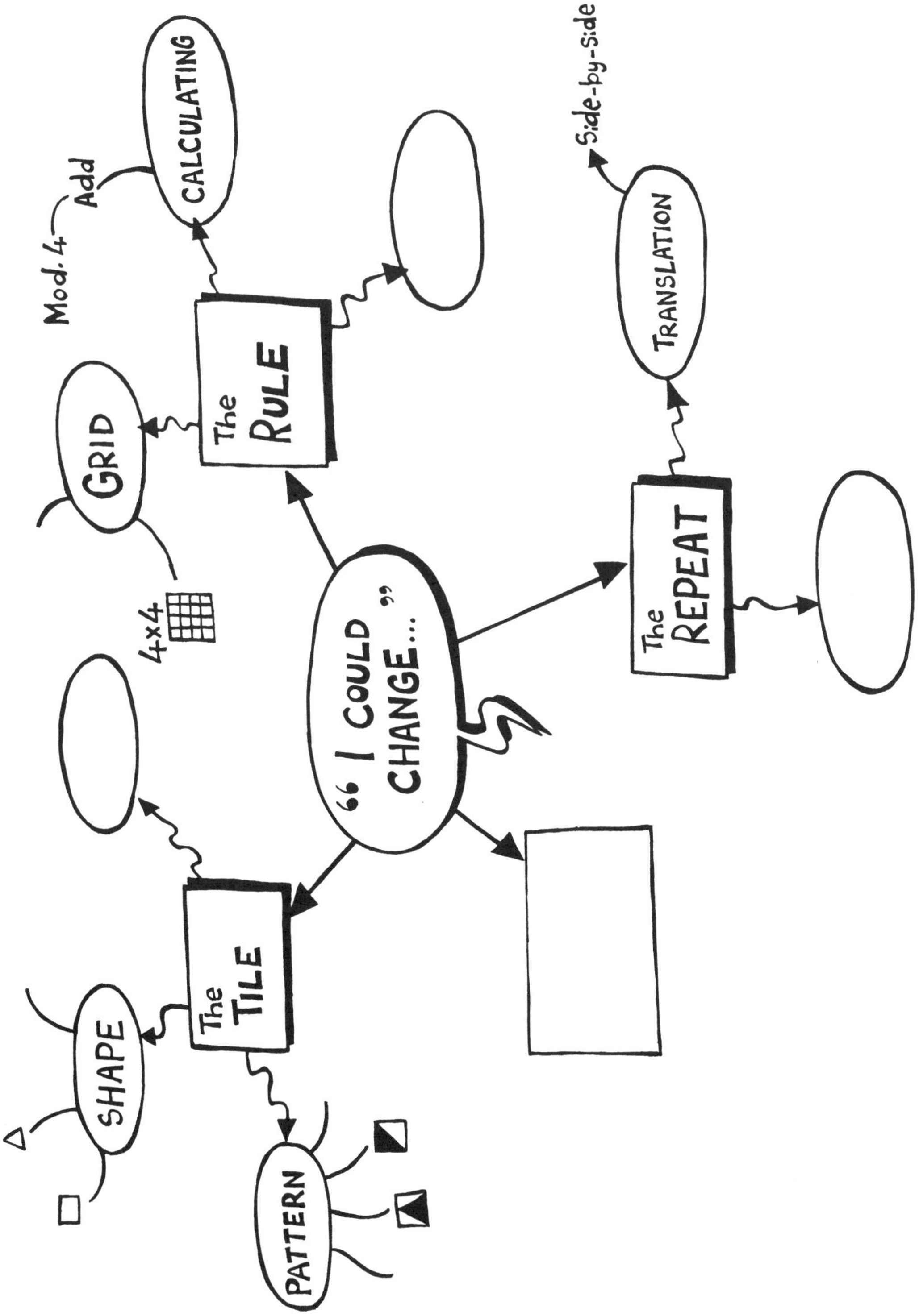

"I COULD CHANGE..."
The RULE
CALCULATING
Add
Mod. 4
GRID
4×4
The REPEAT
TRANSLATION
Side-by-Side
The TILE
SHAPE
PATTERN

At sixes and sevens

'At sixes and sevens' requires the children to display given numbers on a calculator by using only a handful of keys. Calculating strategies will readily spring to mind for some target numbers – no more than a basic knowledge of number bonds is needed to ensure that – but genuinely creative thought is essential to track down other solutions. Some suppose that creativity is at work when answers spring from nowhere but this activity suggests otherwise. By extending and adapting simple starting ideas children find that they can create new solutions that were initially far from obvious.

Key maths links

- Understanding addition and subtraction
- Understanding multiplication and division

Thinking skills

- Creative thinking skills
- Seeing links
- Inventing different solutions

Language

subtract, take away, decrease, multiply, multiple, How many more to make ...?, How about ...? Try ...

Resources

PCM 15 (one per pair)
PCM 16 (optional extension)
calculators

1 Setting the scene

Seat the children in pairs with one calculator shared between them. Display the following numbers and operations on a board or flipchart:

=	3	4	×	−

Challenge the children to make their calculator display the number 1 by pressing only the keys shown (the legal keys). This is simple (), so quickly challenge children to display another number, say 11. The children may use the keys as often as they like, so   is a perfectly valid solution for this target number. Repeat with other target numbers. Remind the children to clear their calculator before each new calculation. As numbers become more challenging compare and discuss the children's different solutions.

2 Getting started

Once the children understand the activity ask them to continue, either using the keys you started with or with a fresh set, e.g. . Ask the children to find ways of displaying each of the numbers from 1 to 12. Use PCM 15 to record the best solutions for each.

Simplify

To make the activity easier, permit the use of one or more additional legal keys. Another operation or another number will simplify the task.

Challenge

Extend the activity to all target numbers up to 20, or 50. Challenge the children to find more than one solution for each target number. Work out the least number of key presses required for each target number. Aim for the most efficient solution. Ask the children to identify unusual or creative solutions.

3 Checkpoints

Encourage experimentation. A few inventive solutions open up the problem since methods that work for one target number can often be adapted for others.

Watch out for ...

In the eagerness to find solutions it is surprisingly easy to forget the rules! Make sure the children are not tempted to slip in the odd $+$ here or there. If they get stuck it is helpful if you have one or two suggestions to prompt them. Spend a few minutes trying the activity yourself beforehand and you'll not be short of ideas.

Ask ...

- ❍ *How can you use this method again?*
- ❍ *Can you think of another way?*
- ❍ *Can you think of a quicker way?*

Listen for ...

Some children will see ways to extend ideas: *That way worked for eight, so I can change it a bit to make nine.* Encourage this approach, as it is both creative and efficient. Though more demanding, some children will try reasoning backwards from the target number. This requires an understanding of inverse operations. It is effective because it takes away much of the guesswork.

 Moving on ...

Review and evaluate the children's methods for finding a range of target numbers. Which target numbers were easiest? Hardest? Which methods proved most adaptable? Which were the most creative?

Where next?

- ❍ Investigate what makes this type of activity difficult or easy. Ask the children to make up similar puzzles of their own with their own choice of legal keys and consider which ones prove easy and which ones prove difficult. Use PCM 16 and record the difficulty of each challenge. Is the choice of starting numbers significant? What about the choice of operations?
- ❍ There are many similar calculator activities. Try '3 hops to 100'. Enter any whole number into a calculator and reach exactly 100 in just three 'hops'. Each hop uses one digit and one operation (e.g. 7 6 [- 1] [÷ 3] [× 4] = 100).

Were you impressed with the ingenuity of some solutions? Did you see efficient methods ripple through the class and become adopted and adapted by neighbouring groups? Or were children defensive about sharing their methods? Perhaps it's okay to pinch others' ideas if you're prepared to adapt them and make them your own? Especially if these new ideas are shared in turn.

DEBRIEF

Name _____ **Date** _____

Can you display the numbers 1 to 12 on a calculator?

Legal keys: ▢ ▢ ▢ ▢ ▢   Key presses

1
2
3
4
5
6
7
8
9
10
11
12

Degree of difficulty

Name _____ Date _____

Try to display the numbers 1 to 12 on your calculator using legal keys of your own.

My legal keys: ☐ ☐ ☐ ☐ ☐

The challenge was:

Very easy	Easy	OK	Hard	Very Hard

My legal keys: ☐ ☐ ☐ ☐ ☐

The challenge was:

Very easy	Easy	OK	Hard	Very Hard

Assessing progress

Assessing the development of creative thinking is challenging as there are often a number of solutions and ideas that can be considered creative in any particular situation. You will have to consider the individual pupil too. A genuinely creative thought for one pupil – something new and insightful for them – may not be so creative in another. Also the process is not necessarily regular or frequent. It is therefore important to consider children's attitudes or their dispositions in different situations. They should be asking questions and be confident to offer ideas. It is this confidence or perhaps playfulness that is the best indicator of creativity, rather than trying to assess specific solutions or outcomes.

Cross-curricular thinking

Literacy

Brainstorming for ideas is a good general technique to develop creative thinking. It is important that it is done in an atmosphere where the children know that offering ideas is more important than coming up with the right answer and where all ideas are accepted uncritically. In literacy this technique can be used when responding to a text to record thoughts and feelings, as well as to stimulate ideas for composition in terms of the content and detail of the vocabulary used. Brainstorming is usually conducted as a whole class activity. It can also be useful to start off in groups so that the children become more independent in using the technique.

Science

Using analogies can be a powerful way to develop scientific understanding in a creative way. Asking children to think of an analogy for something (such as an electric current being like water pipes with the current flowing round a circuit) not only provides an opportunity to compare why they are alike and how they are not alike, but also offers an insight into children's thinking about the science involved.

Design Technology

Coming up with ideas is an essential part of the design process. One technique that can help is to ask children to visualize how the product will be used. Ask them to 'see' it once it is finished: *What will it need to do?*, *What will make your idea different or special?* This can be the basis for more structured planning and development, though the whole process is a creative one.

Geography

The strategy 'Banned' (see page 23) can easily be developed in other subjects and provides opportunities to develop specific vocabulary. However, it can also be a way to stimulate creative thinking as children will come up with imaginative ways to give clues to words, such as: *It sounds like fountain but starts with an 'm'.* They may find it hard to formulate rules for banned ideas or words. It is usually best to praise their ingenuity and finish with a discussion of creative ways to get round the rules.

In my opinion ...
Evaluation skills

> **Evaluation** – these skills enable pupils to evaluate information, to judge the value of what they read, hear and do, to develop criteria for judging the value of their own and others' work or ideas and to have confidence in their judgements. (QCA 2000)

Overview

Evaluation is about taking responsibility for your own opinions and judgements and being prepared to explain or defend them to others with reasons. It requires confidence in knowing what you think and sensitivity in evaluating or criticizing the work of others. It requires the ability to set and apply criteria to tasks and make judgements based on those criteria. The final stage is presenting these judgements to others and being prepared to defend or change that judgement in the light of feedback. This involves awareness of the feelings of others in giving and receiving feedback - a challenging aspect of effective collaboration and an important aspect of speaking and listening.

In mathematics evaluation is essential in developing confidence in knowing that you have a good solution and understanding *why*. Mathematics is often perceived as being about applying rules or being able to remember facts and formulas, however an essential part of being able to think mathematically is to be able to make judgements about which facts to use or which formula to apply. A good solution in mathematics might be an efficient one, or an elegant one, or one that leads to new insights and thinking. Deciding which is the best way to do something mathematically therefore, often calls for evaluation and judgement.

Strategies

Evaluation skills can be broken down further into the following kinds of behaviours or activities that pupils can do:
- **Evaluate information**
  Appraise, assess, critique, decide
- **Judge the value of what they read, hear and do**
  Review, weigh up, scrutinize
- **Develop criteria for judging the value of their own and others' work or ideas**
  Evaluate, judge, mark
- **Being confident in their judgements**
  Express opinions, disagree, agree (with reasons), resolve

Questions

How could you justify that? What reasons are important? Can you explain ...? How will you check it? Can you argue the opposite? Do you agree? Do you disagree? Which do you think?

So what?

BRIEF

When interpreting data presented graphically the repetition of one simple question will encourage evaluation skills ... *So what?* This activity requires children to identify significant facts and consider what they mean. Following a discussion the children make statements that demonstrate understanding or propose an action (or both). The end product is a simple poster displaying graphs and statements together.

Key maths links

- Organizing and using data
- Fractions

Thinking skills

- Evaluation skills
- Distinguishing fact from opinion
- Judging the reliability of evidence
- Making decisions informed by evidence

Language

questionnaire, bar chart, pie chart, fraction, percentage, recommendation, therefore, so

Resources

PCM17 (one per pair/group, possibly enlarged)
PCM 18 (as needed)
coloured backing paper (A3 or larger)
scissors
glue sticks

 ## Setting the scene

Introduce the context for the data on PCM 17: *A class of children designed and completed a questionnaire about homework. These graphs show answers to some of their questions.* Display or distribute graph A. Ask the children to state facts drawn from the graph, e.g. 3 children out of 25 said they do not enjoy homework. Ask: *So what?* Discuss what the stated fact might mean (look at the context of the whole graph). Agree a sentence that sums up a clear understanding of the graph, or proposes a suitable course of action. If possible combine the two in one: *Most children are at least reasonably happy with their homework so their teacher should certainly continue to set it,* or *There are more children who do not like homework than those who enjoy it, so it needs to be changed.* Copy and complete the diagram from PCM 18 on a board to illustrate the thinking processes involved.

Getting started

Supply pairs or small groups with PCM 17 and PCM 18 plus a large sheet of coloured paper to act as a backing sheet. Ask the children to find facts and then interpret them. Encourage them to identify descriptive facts first (*No children dislike homework a lot*) which can simply be 'read' from the graph, then comparative statements (*More children do not enjoy homework than enjoy it*).
Ask the children to consider the significance of each fact by discussing *So what?* Encourage speculation and expect the children to back up their statements to each other. Ask them to record their thoughts on PCM18 and then, based on what they put in the 'Understanding' and 'Action' bubbles, write statements on small sheets of paper. Display these on the backing paper alongside appropriate graphs and add a title and possibly an introductory paragraph and concluding summary.

Simplify

Supply fewer graphs, perhaps one at a time. If using your own data/graphs: make questionnaire options simpler (e.g. Agree, Disagree, Don't know).

Challenge

Require children to combine data to make more persuasive comments: *22 children either agreed or strongly agreed that homework helps them do well in school.* Require children to calculate percentages for their comments: *88% of children thought that ...*

 ## Checkpoints

Help the children select significant information and then exercise their imagination as they experiment with possible meanings. This is not easy. As they begin to understand

how to answer the *So what?* question actively encourage their contributions. To start with, give the 'thumbs-up' to *all* suggestions. When you have established a climate where suggestions are forthcoming *then* it is possible to begin to criticize them, asking if the words and the facts agree or if the proposed course of action fits the situation.

Watch out for ...

Some children will find it very hard to interpret the facts. They may resort to their own opinions, especially when they conflict with views depicted in the charts. Be understanding of this. Encourage a more general view by asking what *most* people think. Some children will try to avoid thinking by concentrating on colouring bar charts and making fancy titles. Set ground rules so the children know what they should concentrate on.

Ask ...

- ❯ *Do your words match the graphs?*
- ❯ *How do you know that?*
- ❯ *So what?*

Listen for ...

Applaud any comments that reflect evaluative thinking: *So ..., Therefore ..., This shows that ..., That means ..., This proves that ..., That's not true, because ....*

 Moving on ...

Ask the children to present one or more of their statements (select groups to prepare this during the lesson). Ask others to find evidence to support their statements. If the children show a good understanding ask them to challenge or contradict statements, always relying on facts to support a counter-case.

Where next?

- ❯ Make your own questionnaires/data – the children can complete some or all of the preparation of the questions and graphs (see 'What's the issue?', Unit 6, Activity 3).
- ❯ Employ particular data-handling skills (e.g. bar charts, line graphs, averages, range, percentages).
- ❯ Employ evaluation skills in any subject where graphs are used, e.g. science, geography.
- ❯ Integrate with the English curriculum to form a persuasive argument: concentrate on an orienting introduction and a concluding summary to place graphs in a clear context.

Did the children make sense of the unfamiliar graphs? Were their interpretations plausible? What topical issues closer to home could be the basis of your own survey?

DEBRIEF

Results of homework questionnaire.

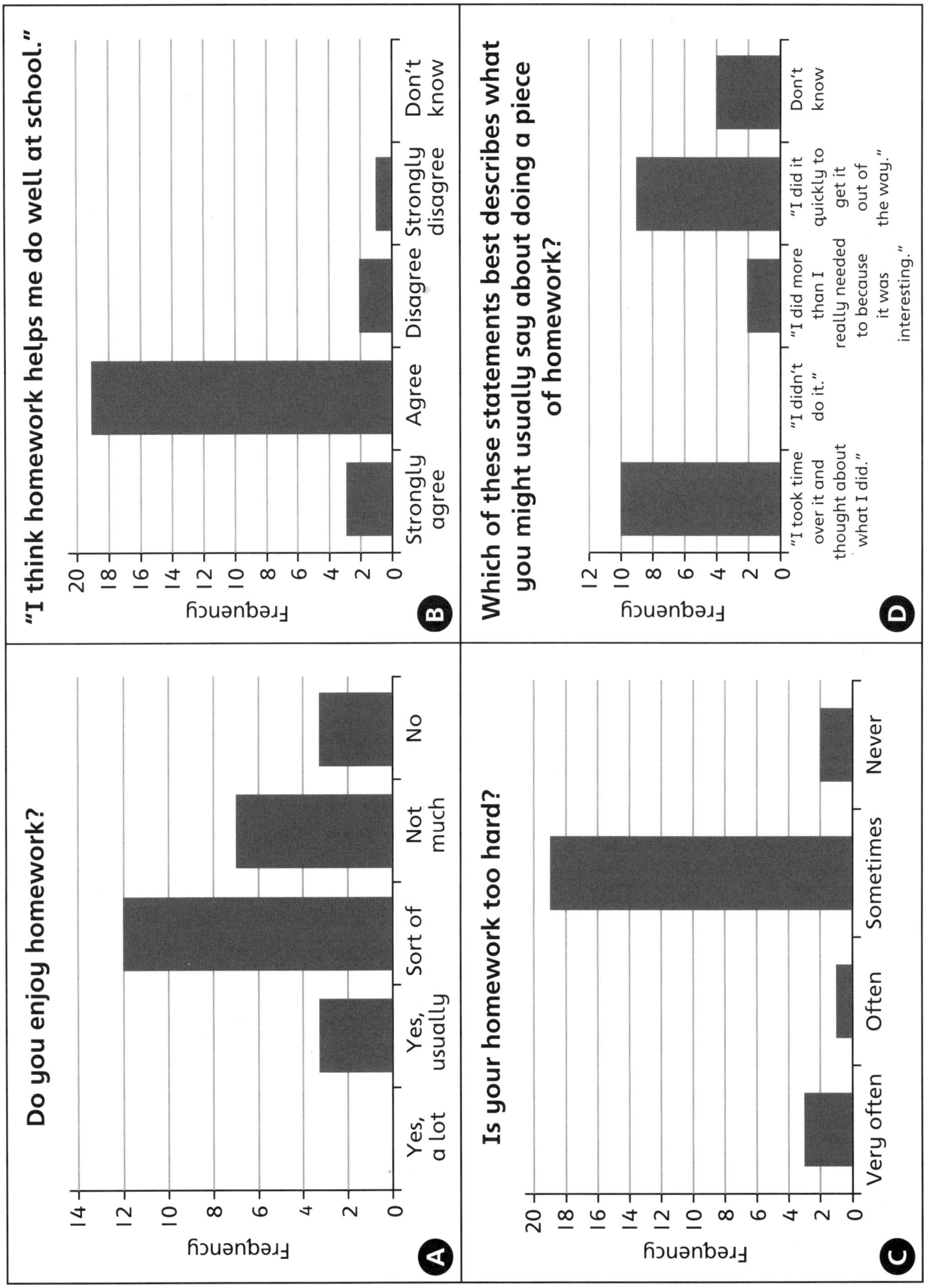

B

"I think homework helps me do well at school."

Frequency (0, 2, 4, 6, 8, 10, 12, 14, 16, 18, 20)

Strongly agree | Agree | Disagree | Strongly disagree | Don't know

D

Which of these statements best describes what you might usually say about doing a piece of homework?

Frequency (0, 2, 4, 6, 8, 10, 12)

"I took time over it and thought about what I did." | "I didn't do it." | "I did more than I really needed to because it was interesting." | "I did it quickly to get it out of the way." | Don't know

A

Do you enjoy homework?

Frequency (0, 2, 4, 6, 8, 10, 12, 14)

Yes, a lot | Yes, usually | Sort of | Not much | No

C

Is your homework too hard?

Frequency (0, 2, 4, 6, 8, 10, 12, 14, 16, 18, 20)

Very often | Often | Sometimes | Never

Finding facts

Name _____ **Date** _____

Find facts from the graphs and write about them.

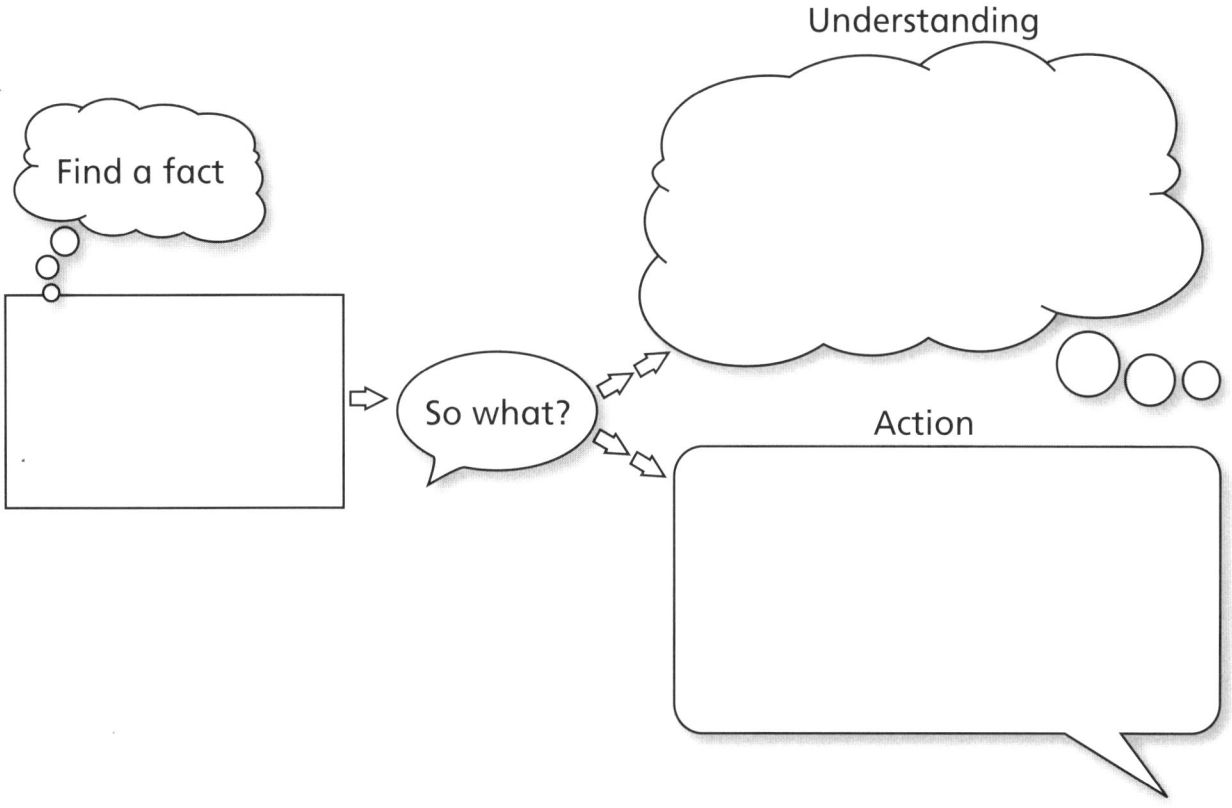

Find a fact

So what?

Understanding

Action

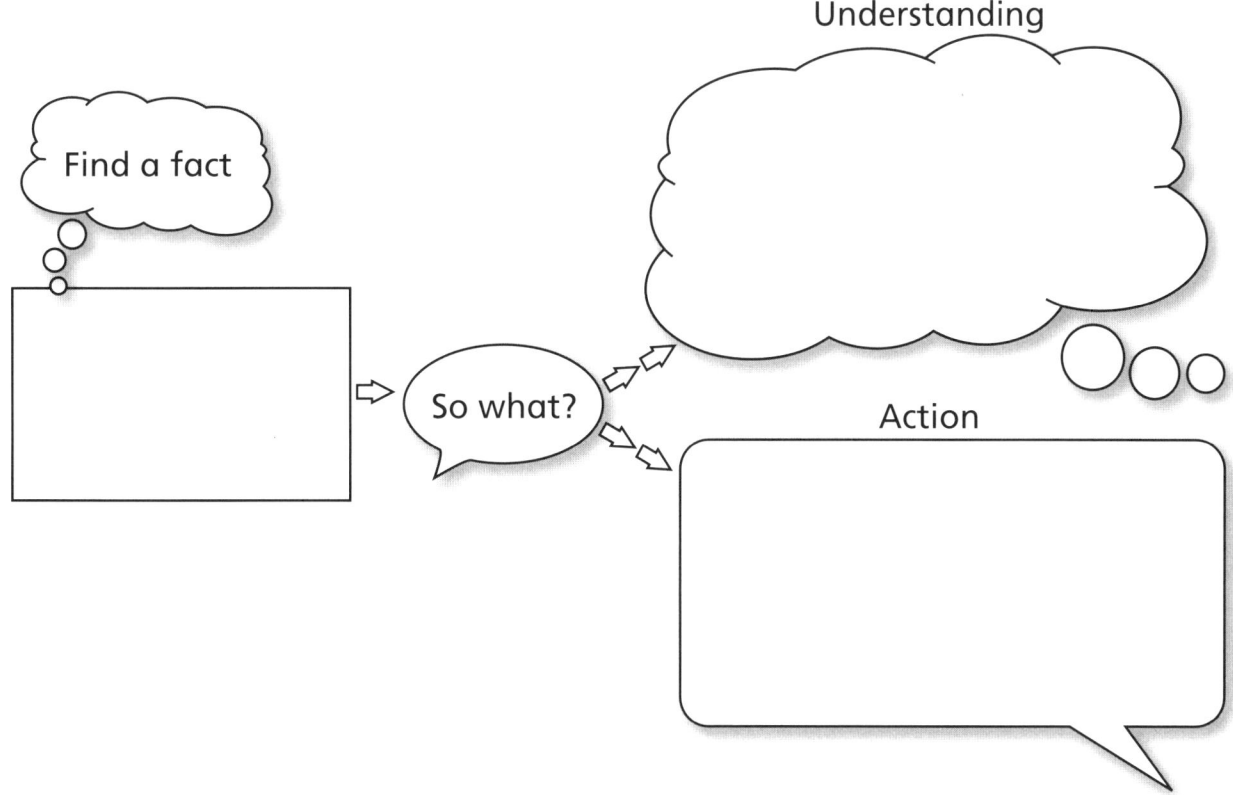

Find a fact

So what?

Understanding

Action

Domino dots

'Domino dots' poses a simple question: *How many are there?* However, as in many real life situations, there is enough opportunity for error to make the question non-trivial. The children are challenged and guided to find a range of methods to answer the question. They then weigh up their methods thoughtfully to decide the best approach, considering carefully what 'best' might mean in this context.

Key maths links

- Mental calculation strategies
- Checking results of calculations

Thinking skills

- Evaluation skills
- Analysing
- Inventing different solutions
- Weighing up pros and cons

Language

addition, sum, total, strategy, Show how …, Explain …

Resources

PCM 19 (one per pair – top part only to start with)
PCM 20 (one per pair, or one copy for display)
dominoes (optional)

Setting the scene

Spread out a full set of double-6 dominoes or display a copy of PCM 19 (top part only). Ask: *How many dots are there?* Consider together how to proceed.

Getting started

Organize the children in pairs. Distribute PCM 19 (top part only). The challenge is to find out how many dots there are. Some may see this as a competition and rush to be the first to find the answer. Errors are likely so you can challenge answers until a consensus emerges. (There are 168 dots.) Ask the children to explain their methods to each other. Review some of the most promising approaches with the class. Encourage different ways of analysing the highly regular pattern of dots: try totalling across the page, totalling down the page, totalling individual rows, or whole dominoes, constantly looking for patterns. Write down the numbers and organize them in different ways: look for patterns.

Simplify

Give pointers where necessary to start children off: *How many ones are there? How many twos …?; Will you add … or multiply?* Use a calculator to help.

Challenge

Expect a range of methods from more able children. Hint that it would be good to find a general method that could be extended to a larger set of dominoes (such as a double-9 set: see PCM 19 (full sheet)). *What is the average number of spots on a domino? Can you simply multiply the average number of spots by the total number of dominoes?*

Checkpoints

The focus is on finding a good method. In conversation with pairs/groups of children consider what makes a method 'good' so that you have prepared the ground for subsequent whole-class discussion. Consider which solution has the fewest calculations.

Watch out for ...

Most children will find the number of dots accurately, given time. Some children may consider the activity complete when this has been achieved. Emphasize that finding the answer is not in fact the main challenge. Encourage creative thinking to discover alternative methods.

Ask ...

- ◑ *Can you explain your method?*
- ◑ *Is it a good method? Why? Why not?*
- ◑ *Can you find a better method?*
- ◑ *Which is the best method? Why?*

Listen for ...

Most children will see regularity in the pattern of dots. Applaud any approach that uses regularity to aid calculation. The need to add consecutive numbers may emerge. Some children will appreciate being shown the following 'trick':

$$1 + 2 + 3 + 4 + 5 + 6 = (1 + 6) + (2 + 5) + (3 + 4) = 7 + 7 + 7 = 7 \times 3$$

Moving on ...

Review the methods that the children have invented. What characteristics do they consider make a 'good' method? (You may need to offer suggestions.) Criteria could include: accuracy, speed, ability to identify errors, cleverness, and the ability to extend the method to other problems. In pairs, or as a whole-class, complete PCM 20 listing methods (can the children invent catchy names to distinguish them?) and desirable characteristics. Evaluate each method by ticking (or scoring) against agreed criteria. Ask: *Do you consider any method to be 'the best'? Can you say why?*

Where next?

- ◑ Challenge the children to find the number of dots in a set of double-9 dominoes (PCM 19 – whole sheet). If the children have found an efficient method they may be able to complete this challenge in a few minutes. The 'trick' shown above can be extended if it is helpful (include 0 to make it work):
  $$0 + 1 + 2 + 3 + 4 + 5 + 6 + 7 + 8 + 9 = (0 + 9) + (1 + 8) + (2 + 7) +$$
  $$(3 + 6) + (4 + 5) = 9 \times 5$$
- ◑ Recall the Christmas song 'The Twelve Days of Christmas'. How many gifts did my true love give to me?
- ◑ Use PCM 20 to evaluate the best way to do something in any context where a range of options is open.

After the initial enthusiasm to find the answer, were children willing to seek 'good' methods? How did you encourage this? Could children identify criteria for judging between their various methods? Could they justify their own preferences? Were they able to criticise others' methods sensitively?

DEBRIEF

How many dots?

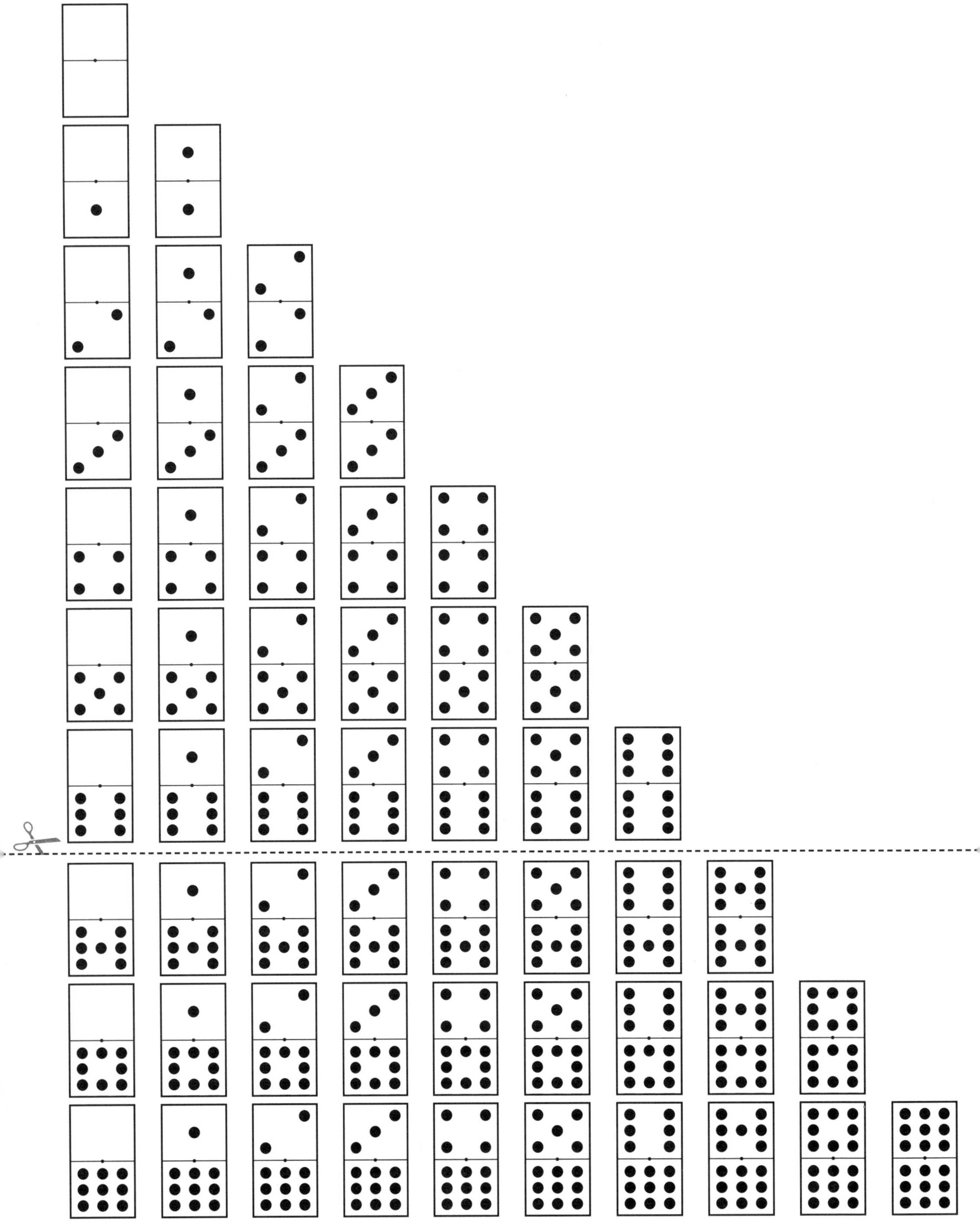

Methods for finding how many

Name _____ **Date** _____

List the things that make a good method here

Give each of your methods a name and write them here

Methods / Characteristics			

Assessing progress

The children's increased confidence in their own thinking is one of the hallmarks of improving evaluation skills. It is about taking responsibility for your own opinions and judgements and being prepared to explain or defend them to others with reasons. This requires confidence in knowing what you think. This confidence should be justified, of course, so children should be prepared to change their minds if necessary, in the light of information or reasoning. At this stage it is also important for children to show sensitivity in evaluating or criticizing the work of others.

Cross-curricular thinking

Literacy

Assessment for learning (see page 14) strategies such as 'Traffic lights' are good starting points to develop evaluation skills. You can ask the children to rate a piece of writing that they have done with green for: *I think I can go on*, orange for: *I think I am getting going*, and red for: *I'm at a full stop here*. This opens up the way to discuss criteria for success in the task so that children can evaluate their own work.

Design technology

Evaluation is also central to design technology. The children need to learn to develop evaluation criteria for their designs in order to guide their thinking as they work. This should be an integral part of the process and not simply a retrospective review. Using a digital camera to record the process of designing and making enables the children to recall what they were thinking at the different stages and reflect on the criteria to evaluate the task.

History

A strategy such as a 'Mystery' (where snippets of information are pieced together to answer a central question) can help children to use their evaluative skills as they judge the importance of the different 'clues' they have been given. In history this can be a good way to assess understanding of what has been learned in a unit of work as they use their historical knowledge to do this. Clues can easily be written to support a discussion about: *Who was responsible for the Great Fire of London?* for example, to get children to see that the baker may have started the fire, but that there are other factors to consider.

Geography

Some other general techniques that are helpful in developing evaluation skills are those developed by Edward deBono where children are given thinking frames with headings such as 'Plus, minus and interesting' (PMI) (see page 23) or a focus on 'Consider all factors' (CAF). The structure of the sheet helps children to think more carefully and give more considered responses. These approaches can be combined with collaborative discussion (such as 'Think, Ink, Pair, Share' where children are asked to consider their response, make some notes, discuss it with a partner then in a group). This can be particularly useful in a subject like geography when the children have to evaluate changes to the environment or express their views about people and places.

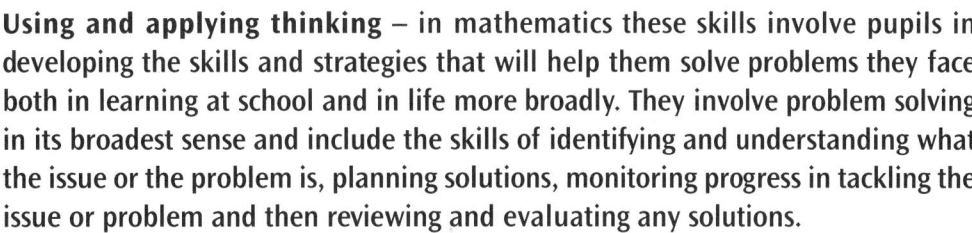

Think on!

Using and applying thinking skills

Using and applying thinking – in mathematics these skills involve pupils in developing the skills and strategies that will help them solve problems they face both in learning at school and in life more broadly. They involve problem solving in its broadest sense and include the skills of identifying and understanding what the issue or the problem is, planning solutions, monitoring progress in tackling the issue or problem and then reviewing and evaluating any solutions.

Overview

The aim of this unit is to identify some activities for pupils to put their mathematical thinking skills into practice. This will give them the opportunity to evaluate how well they have developed their skills through the earlier activities as well as giving you the opportunity to assess how well they can apply what they have learned. The activities are set as challenges, problems or puzzles.

The process of undertaking these activities relates to the different kinds of thinking in the earlier units. The early stages draw on information processing skills by focusing the children on what they have to do and what they already know. There may be scope for creativity in seeing alternatives or applying knowledge and skills imaginatively to a new problem. Enquiry skills are brought into play during the main part of the activity as any solution is formulated and tested, closely supported by reasoning skills which also help to link the different stages and ensure continuity throughout the process. Evaluation skills are essential to appraise and review any solution and to develop confidence in being successful.

Strategies

Supporting the pupils in using and applying thinking skills to problems is best framed as a series of questions:

- **What do we have to do?**
  What is the problem, challenge or issue to be resolved?
- **Where do we start?**
  What do we know?
  Have we done anything like this before?
  What possibilities are there?
- **How will we know when we have got there?**
  What will a successful solution look like?
- **Are we on track?**
  Is this going to lead us to the answer we imagined?
- **Have we got there?**
  Is this a solution to the problem we were set?
  Could we have done it differently? Is it the best solution?

Questions

What do you have to do? What do you need to know? What do you know already? Have you seen anything like this before? What could you try? Do you think that will work? What will the answer look like? How could you test that? How can you check that? Is this the best answer? How else could you have done it?

Game on

BRIEF

Asking children to design and make a game is an excellent way for them to apply many skills. Setting up the conditions so that the final product is in each case impressive is your principal challenge as teacher. This requires the children to have a large reservoir of starting ideas, insight to select the most promising ones, confidence to develop them … and plenty of time. This activity is unlikely to be a single lesson.

Key maths links

○ Dictated by you or the children

Thinking skills

○ Using and applying skills
○ Developing ideas
○ Applying imagination
○ Evaluating

Language

mathematical vocabulary dictated by you or the children

Resources

PCM 21 (one per group, copied on card)
PCM 22 (optional, one per group, copied on card)
other equipment chosen by the children
scissors
counters

Setting the scene

If you have them, display a range of table-top games. If possible, play some of them. Ask the children to name as many similar games as possible. Create a list. Together, categorize games on the list: *Games where playing pieces are captured*, *Games where players race to a finish*, etc. Invite the children to create their own mathematical game, based on the given board (PCM 21). In the light of your introductory discussion about familiar games, consider with the children good features to incorporate in their own, e.g. a clear aim, a suitable scoring system, specific equipment, unambiguous rules … Emphasize that their games must include mathematical content. This could be in the way players gain points, in the need for mathematical knowledge, in the use of reasoning skills or in any other way. If you want to, dictate the content, e.g. requiring multiplication or division skills. This is a good way to reinforce recent work.

Getting started

Working in small groups the children design and make a game for two or more players. Games can use any easily available equipment in addition to the board itself, which can be decorated or annotated in any way. If needed, game pieces to suit the board are provided on PCM 22. Other useful items include blank cards, standard dice, blank dice, and calculators. Focus the children's attention firmly on the way the game is played. Promise a future opportunity to create smart game boards, fancy playing pieces etc. (if you are prepared to find the time by extending the task into design technology). Ask them to present the key features of their game to the class to get feedback.

Simplify

Stipulate that the game must practise specific skills, e.g. knowledge of multiplication/division facts or stipulate that the game must be suitable for younger children and make explicit the mathematical content that is appropriate.

Challenge

If you know the children's mathematical skills are not being applied in the game that they are developing then challenge them to include more demanding ideas. Be a critical friend and offer suggestions that will extend their initial ideas.

Checkpoints

Watch out for ...

Some groups will create simplistic games. This may be because the children concerned have had only limited experience of playing board games. Suggest a few twists to the simple ideas proposed, but try not to impose your own. You will not be able to make up for a lack of experience in a short time.

Ask ...

- ◗ *What is the aim of your game?*
- ◗ *What are your rules?*
- ◗ *How can you improve the game?*
- ◗ *Is the game fun to play?*

Listen for ...

Be on the look out for examples of children taking a familiar idea and adapting it to suit their own designs: *That's how it works in my game at home, but we'll need to change it a bit ...*

Moving on ...

The creation of a good game can be very time-consuming. The children are unlikely to invent the very best rules straightaway. Find time for them to trial their games, improve them and play them with others who were not involved in their design. You could find time to improve the presentation of the games and create clear written instructions. Of course, these are not obviously mathematical tasks. Devote as much time as you see fit to these refinements. Consider with the class the thinking and mathematical skills needed to make the games. List them.

Where next?

- ◗ Improve presentation of game boards and playing pieces.
- ◗ Make decorated boxes to contain the games.
- ◗ Write clear rules.
- ◗ Create advertising for the games.
- ◗ For a quicker activity, design simple calculator games requiring no more than pencil and paper and a calculator to play.

A good test of an open activity such as this is the amount of variety shown in the final products. If groups have produced a wide range of outcomes it means that they began with a fund of ideas, knew how to develop them and had the confidence to do so. Were you pleased with the games? Were they sufficiently mathematical? What could you do another time to make the outcome even more impressive?

DEBRIEF

PCM 21

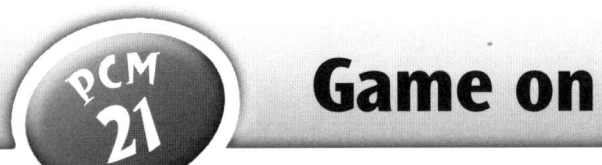

Game on

Name ______________________________ Date ________________

Design a game to be played on this board.

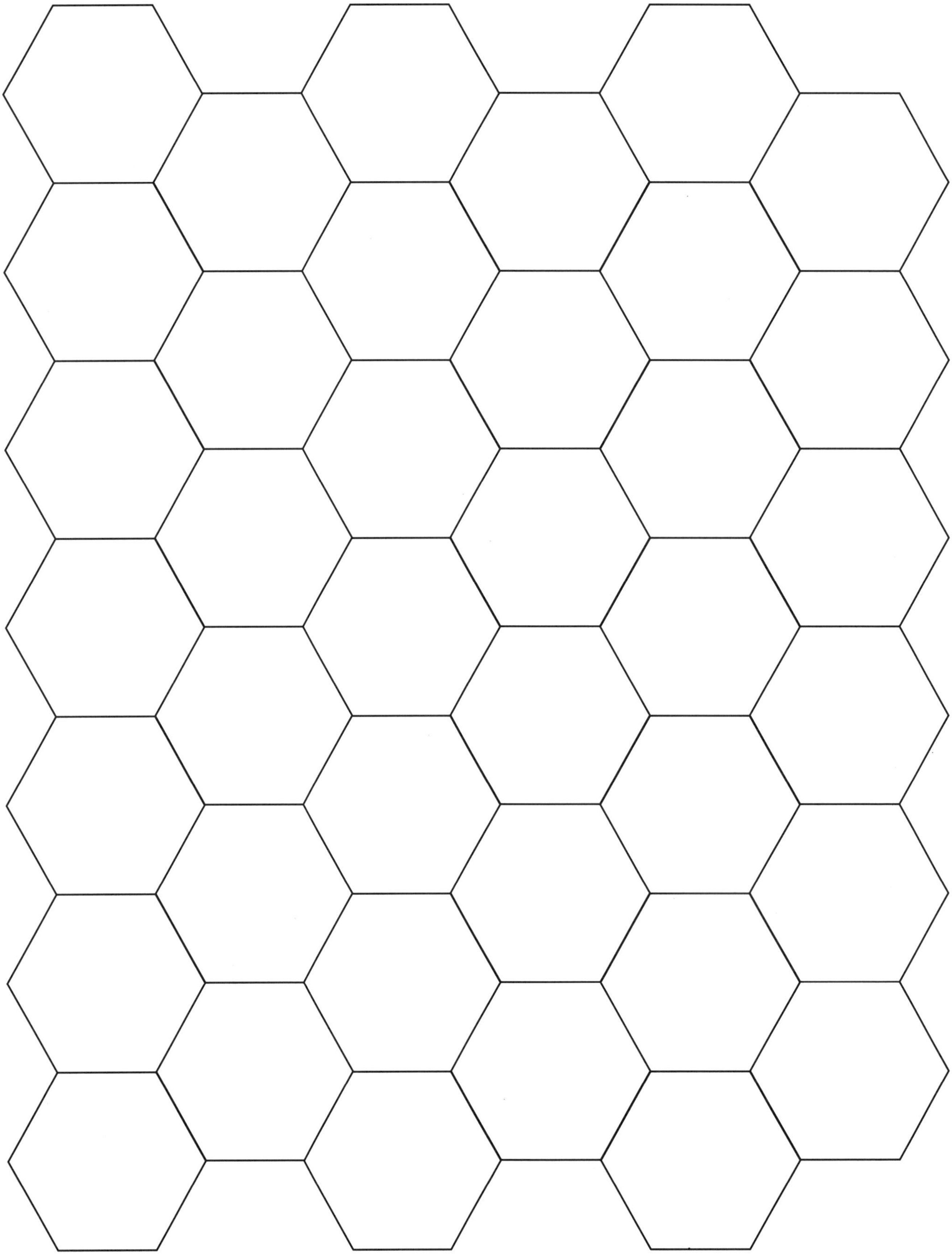

Game players

Cut out and use these as game pieces.

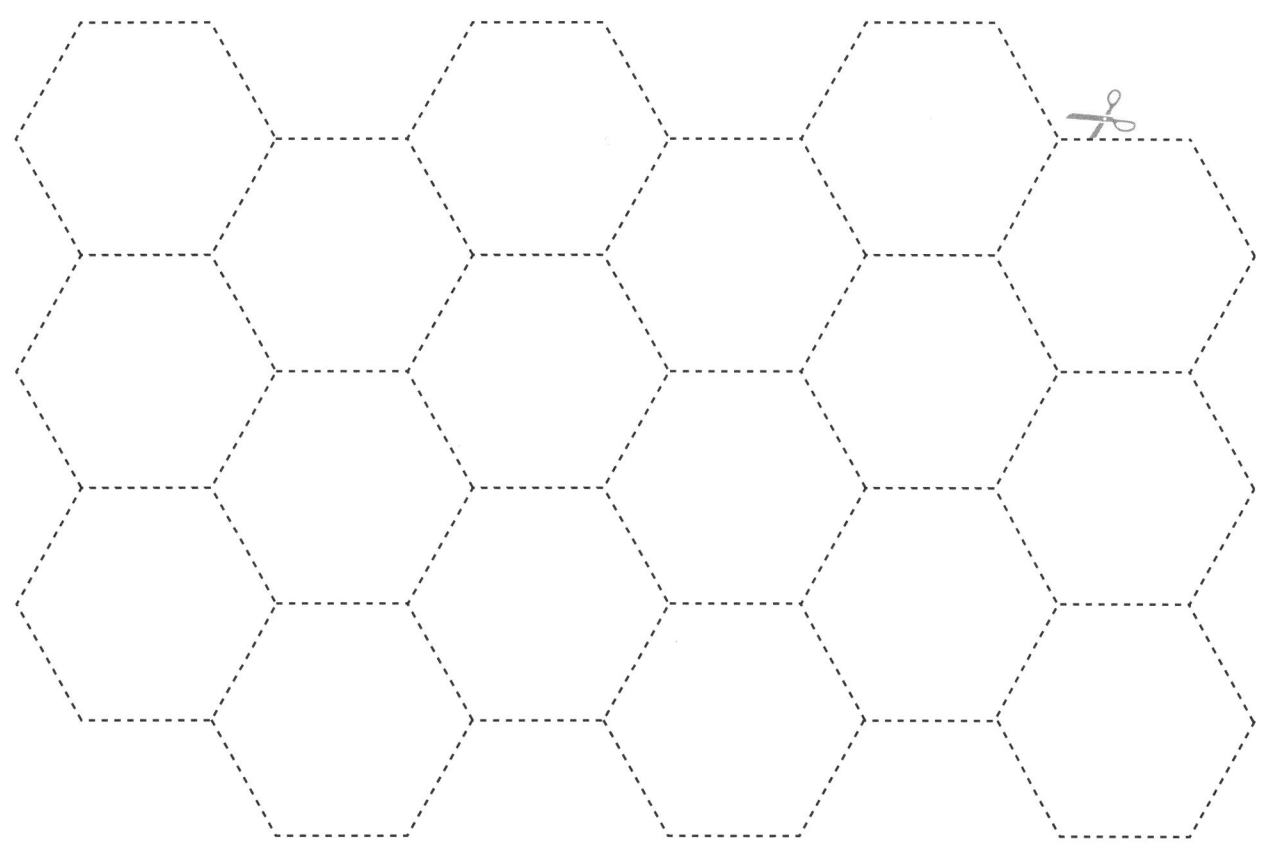

Classroom reshuffle

Productive thinking often involves working with others. In 'Classroom reshuffle' the children combine thinking skills and practical measurement with teamwork to improve the classroom environment. The activity may take two or more lessons.

Key maths links

- Measures
- Fractions

Thinking skills

- Using and applying skills
- Problem solving
- Weighing up pros and cons
- Decision making
- Giving reasons for opinions

Language

proportion, scale, area, square metre

Resources

PCM 21 (one per group, preferably on coloured card)
PCM 22 (one per group)
tape measures
metre sticks
calculators
glue sticks

 ### Setting the scene

With your class consider the physical arrangement of the classroom. Identify good and bad features. List the features of a well-organized classroom on a board or flipchart. Consider how to plan improvements. Identify the need for a scale plan of the room and its contents to allow experimentation. Clarify the meaning of 'scale' and demonstrate the way to scale down from full-size measurements to a scale drawing. Establish those dimensions needed for a plan view, i.e. length and width but not height. Consider a suitable scale for drawing the classroom and agree what it will be for this activity (1:50 will be about right).

Getting started

Organize the children in threes or fours. Distribute PCM 23. Ask the groups to gather dimensions of the classroom furniture etc. then convert these to scale values and make scale drawings on the mm-square grids to cut out and arrange on a floor plan. Making the plan of the room could be part of the activity or you could draw it (on cm-square grid paper showing all fixed features, e.g. windows) and supply copies. Then the children can arrange the movable components carefully, taking account of the list of good and bad features. Encourage discussion and repeated rearrangement of furniture. Give this part of the lesson plenty of time. Put items in place then, when satisfied, glue them in their final position. Each group can complete its proposal by mounting its plan on a coloured backing sheet, adding annotations to explain and promote its best features.

Simplify

Supply the floor plan for some or all groups. Have master copies of the room's furniture drawn to scale - supply these to augment the children's efforts.

Challenge

Ask specific groups to undertake the challenge of drawing the floor plan to scale. Later, suggest groups code their completed plan according to use (e.g. seating, library, craft, resources). Calculate the area devoted to each activity (in m^2).

 ### Checkpoints

Give the children the chance to experience difficulties and solve their own problems. If preparing the scale drawings takes a long time and you judge that concentration is waning then stop the lesson. Save the arrangement of the room and subsequent discussion for another day. Distribute envelopes or plastic wallets for groups to store their components in.

Watch out for ...

Ensure all measuring is completed accurately. It's surprisingly easy to line up the wrong end of a metre stick, reading 40 cm for 60 cm, for example. You will be able to spot if drawings are not to scale: groups must be able to trust their drawings to be able to complete the activity. When locating furniture on the floor plan take account of windows.

Ask ...

- ❍ *What is your particular role in the group?*
- ❍ *Tell me an example of good teamwork from your group.*
- ❍ *Why have you suggested this arrangement? Do you all agree?*

Listen for ...

Some groups will organize themselves well, distributing tasks efficiently: *You two tell me the measurements and I'll use the calculator to work out the scale sizes.* Some groups will monitor their own work and notice if things go awry: *There's no way that's the right size - calculate it again.* Some children will offer clearly expressed opinions about their proposals and have good reasons for their decisions. Some children will listen attentively to opinions within their group, asking intelligent questions, seeking clarity and agreement.

 Moving on ...

Involve the children in evaluating the proposals. Place plans and accompanying annotations on desktops alongside comment sheets (PCM 24). The groups move around the room looking at proposals, noting good features (✓) and those that they are unsure of (?) in the appropriate rows. Establish ground rules in advance so that the children's comments are supportive (e.g. *All comments should be supported with a reason; For every ? you must also add a ✓; Don't simply repeat what others have said ...*). Consider explicitly the teamwork displayed throughout this activity. Celebrate successes. Be specific. If desired, ask the children to identify the kinds of thinking that they have used, such as creativity or reasoning (see PCM 26). Finally, review proposals, identify improvements and then move the furniture!

Where next?

- ❍ Place a physical coordinate grid of strings over the school's wildlife area. Map the region to scale.
- ❍ Apply the concept of scale in map work and field studies in geography.

This activity requires practical measuring skills, critical thinking skills and the ability to work collaboratively. Did your management of this activity allow children to exercise all three? In which area did you see the best performance? What will you do to enhance this and the other areas in future?

DEBRIEF

Classroom furniture

Make scale drawings of the classroom furniture.

Scale is ▢

Desks

Dimensions

Full-size:	Scale:

Quantity:

Item:

Dimensions

Full-size:	Scale:

Quantity:

Item:

Dimensions

Full-size:	Scale:

Quantity:

Item:

Dimensions

Full-size:	Scale:

Quantity:

Item:

Dimensions

Full-size:	Scale:

Quantity:

Item:

Dimensions

Full-size:	Scale:

Quantity:

Item:

Dimensions

Full-size:	Scale:

Quantity:

Item:

Dimensions

Full-size:	Scale:

Quantity:

Thinking by Numbers 5 • Unit 6: Think on! • Using and applying thinking skills

Classroom reshuffle

Name _____ Date _____

Rate the success of the classroom reshuffle by writing your views:
good (✓) and uncertain (?).

✓
✓
✓
✓
✓
✓

?
?
?

What's the issue?

BRIEF

This activity is about gathering data for a purpose. Don't impose it artificially but, rather, be alert to situations where there is a need for a survey or questionnaire to help shed light on an issue. This may arise in other studies (e.g. geography or design and technology). Or there may be a real issue affecting pupils that arises out of class or school council discussions (e.g. homework, playtime snacks, cycling to school). Capitalize on the interest generated by such issues, but realize you may need to act on what you find!

Key maths links

- Organizing and using data
- Problems involving 'real life'

Thinking skills

- Using and applying
- Collecting relevant information
- Defining the problem
- Generating ideas
- Distinguishing fact from opinion

Language

questionnaire, survey, tally, frequency

Resources

PCM 25 (one per group)
PCM 26 (for display or one per child)

Setting the scene

Consider the issue that has prompted a need for collecting data. Use structured discussions, role-play, hot seat interviews and other techniques to engage the children's interest. It is important to air the underlying issues and help the children see beyond their personal thoughts and opinions. If the questionnaire is to include opinions and not facts alone children must appreciate the possible range of responses that might be forthcoming. When you judge it appropriate, narrow the discussion down to identify several areas where gathering data may be useful. List these. Take one issue and model how this can be formulated as a clear question. Finally explain that multiple choices must be offered to respondents if analysis is to include the use of graphs and charts. Together, work out some possible answers. Attempt to create a list that is both short and gives room for a variety of replies.

Getting started

Organize the children in groups of three or four. Distribute copies of PCM 25. The groups can copy the processes previously modelled to generate a range of questions using the list of issues, made earlier, as a prompt.

Simplify

Some children will find it hard to put themselves in others' shoes and identify alternative responses to their own. If you can, form groups where there is likely to be a range of opinion.

Challenge

There is considerable scope for independence in this activity. Highly confident children will enjoy being given freedom to invent their own questions, prompted by issues that they have identified themselves. Play devil's advocate and suggest contrary or outlandish opinions.

Checkpoints

Be actively engaged in the groups' discussions. If you judge that the children's understanding and experience is limited then stop the activity. Take responsibility for this: *I can see I haven't given you many chances to see a questionnaire in action.* Rectify this by showing commercial questionnaires (schools receive lots of these) or by using the activity 'So what?' (Unit 5, Activity 1). Consider doing this anyway if you have doubts about your class's level of experience. Children need prior experience to know that they can resort to a range of standard responses such as *Strongly agree, Agree, Disagree, Strongly disagree.* Teach them the need for the *Don't know* response (or its equivalent), to ensure that everyone can reply to the question.

Watch out for ...

Ensure that questions promise to give useful information. Ensure that multiple choices cover as wide a range of opinion as possible.

Ask ...

- *What do you want to find out?*
- *What would be a good question to ask?*
- *Can you predict the answers you'll get?*

Listen for ...

Spot examples of questions being framed with the original issue in mind, and applaud them. Children who relate their work back to the purpose of the investigation will frame better questions and generate more useful data than those who don't.

 Moving on ...

At the end of the lesson share several examples of good questions and multiple-choice answers with the whole class. Explain why you have selected examples for particular attention. Gather everyone's work. Later, compile a questionnaire based on the class's best suggestions. Then distribute the questionnaires, gather the responses and prepare them for analysis. If you want the children to undertake these administrative tasks then use only limited amounts of data to make the task manageable. Alternatively take personal responsibility for preparing graphs and charts that adequately represent the real life issue, based on the children's data, and let them exercise their creativity by working out what they all mean (see 'So what?' (Unit 5, Activity 1)).

Where next?

- Remain vigilant for opportunities to apply data handling skills in a range of settings, especially on real issues in school.

If you undertake a complete survey with your class – from design stage, to data gathering and analysis, right through to making recommendations – then you will have given the children a chance to apply the majority of the thinking skills that they need. You can use PCM 26 as a checklist to discuss this with the children and for them to identify how they have undertaken the different kinds of thinking.

DEBRIEF

PCM 25

What's the issue?

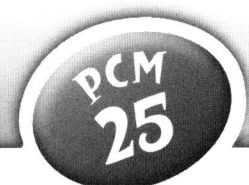

Name ____________________________ Date ________________

Use the template to design your questionnaire.

Issue:

Question:

Possible replies:

•
•
•
•
•

Issue:

Question:

Possible replies:

•
•
•
•
•

Thinking about the issue

Name _____ Date _____

Use this as a checklist when you have completed your questionnaire!

DREAMING & INVENTING
- ☐ Having promising ideas
- ☐ Taking good ideas further
- ☐ Making up theories that might explain things
- ☐ Using my imagination to dream up new things

KNOWING WHAT'S WHAT
- ☐ Recognizing what's worthwhile
- ☐ Weighing up pros and cons
- ☐ Telling facts from opinions
- ☐ Knowing what is trustworthy
- ☐ Knowing what I think

ASKING QUESTIONS
- ☐ Knowing what questions to ask and …
- ☐ Knowing how to ask them
- ☐ Planning ahead
- ☐ Having an idea of what to expect
- ☐ Improving ideas

THINKING SKILLS

MAKING LINKS
- ☐ Seeing that one thing leads to another
- ☐ Working out exactly what I mean
- ☐ Giving reasons for what I think and do
- ☐ Explaining myself clearly

DEALING WITH THE FACTS
- ☐ Working out the information I need
- ☐ Gathering the information I need
- ☐ Organizing the information I've gathered
- ☐ Investigating the information I've organized

Assessing progress

The aim of this final unit was to offer some activities for pupils to put their mathematical thinking skills into practice. This should have given you the opportunity to evaluate how well they have developed their skills through the earlier activities as well as the opportunity to assess how well they can apply what they have learned. The grid below is a way for you to review where you think the children have made progress. It is designed for you to use on the whole class, but could be used to reflect on individual children. It is set out as a grid so that you can indicate where you think the first five units were successful, whether the children were able to show these skills in the activities in Unit 6, where you think you have seen progress in other areas of the curriculum, and where you think the children have developed their awareness of their thinking skills. You may wish to review the activities with a colleague who has also been using the *Thinking by Numbers* activities.

Thinking skills		Units 1–5	Unit 6, Using and applying	Across the curriculum	Awareness of the skills
Information processing	locate and collect relevant information				
	sort				
	classify				
	sequence				
	compare and contrast				
	analyse part/whole relationships				
Reasoning	give reasons for opinions and actions				
	draw inferences				
	make deductions				
	use precise language to explain what they think				
	make judgements and decisions informed by reasons or evidence				
Enquiry	ask relevant questions				
	pose and define problems				
	plan what to do and how to research				
	predict outcomes and anticipate consequences				
	test conclusions				
	improve ideas				
Creative thinking	generate and extend ideas				
	suggest hypotheses, to apply imagination				
	look for alternative innovative outcomes				
Evaluation	evaluate information				
	judge the value of what they read, hear and do				
	develop criteria for judging the value of their own and others' work or ideas				
	have confidence in their judgements				

Appendix

Scope and sequence chart

Unit	Unit name	Activity name	Key maths links	Thinking skills	Page no.
1	Sort it out! *Information processing skills*	Easy as pi	• Measures • Problems involving measures	• Information processing skills • Planning what to do • Comparing	26–29
		Who wants to be a millionaire?	• Estimating and rounding • Problems involving 'real life' and money	• Information processing skills • Planning what to do • Collecting relevant information • Having confidence in judgements	30–33
2	That's because … *Reasoning skills*	Chains	• Properties of number sequences • Rapid recall of addition facts	• Reasoning skills • Ordering information • Explaining • Giving reasons for conclusions	36–39
		Braille and beyond	• Shape and space • Reasoning about shapes	• Reasoning skills • Working systematically	40–43
3	Detective work *Enquiry skills*	Lost in the loop	• Mental calculation strategies • Rapid recall of multiplication facts	• Enquiry skills • Ordering information • Giving reasons for conclusion	46–49
		Fenced in	• Measures • Shape and space	• Enquiry skills • Suggesting hypotheses • Experimenting • Refining	50–53
4	What if …? *Creative thinking skills*	Tiling to order	• Shape and space	• Creative thinking skills • Applying imagination • Generating ideas • Extending ideas	56–59
		At sixes and sevens	• Understanding addition and subtraction • Understanding multiplication and division	• Creative thinking skills • Seeing links • Inventing different solutions	60–63
5	In my opinion … *Evaluation skills*	So what?	• Organizing and using data • Fractions	• Evaluation skills • Distinguishing fact from opinion • Judging the reliability of evidence • Making decisions informed by evidence	66–69
		Domino dots	• Mental calculation strategies • Checking results of calculations	• Evaluation skills • Analysing • Inventing different solutions • Weighing up pros and cons	70–73
6	Think on! *Using and applying thinking skills*	Game on	• *Dictated by you or children*	• Using and applying skills • Developing ideas • Applying imagination • Evaluating	76–79
		Classroom reshuffle	• Measures • Fractions	• Using and applying skills • Problem solving • Weighing up pros and cons • Decision making • Giving reasons for opinions	80–83
		What's the issue?	• Organizing and using data • Problems involving 'real life'	• Using and applying • Collecting relevant information • Defining the problem • Generating ideas • Distinguishing fact from opinion	84–87

Thinking by Numbers 5 and the NNS Unit Plans

The following chart shows how the thinking activities could be used if following the teaching order suggested in the NNS Unit Plans. Choose an appropriate activity to suit your class.

Autumn Term				
		Thinking by Numbers		
Unit	**Unit topic**	**Activity name**	**Thinking skill**	**Page no.**
1	Place value	Unit 1: Who wants to be a millionaire?	Information processing	30–33
2	Multiplication and division 1	Unit 3: Lost in the loop	Enquiry	46–49
		Unit 4: At sixes and sevens	Creative thinking	60–63
		Unit 5: Domino dots	Evaluation	70–73
3	Multiplication and division 2	Unit 3: Lost in the loop	Enquiry	46–49
		Unit 4: At sixes and sevens	Creative thinking	60–63
		Unit 5: Domino dots	Evaluation	70–73
4	Fractions	Unit 5: So what?	Evaluation	66–69
		Unit 6: Classroom reshuffle	Using and applying	80–83
5	Fractions, decimals, percentages, ratio and proportion	Unit 5: So what?	Evaluation	66–69
		Unit 6: Classroom reshuffle	Using and applying	80–83
6a	Handling data 1	Unit 5: So what?	Evaluation	66–69
		Unit 6: What's the issue?	Using and applying	84–87
6b	Handling data 2	Unit 5: So what?	Evaluation	66–69
		Unit 6: What's the issue?	Using and applying	84–87
7	**Assess and review**			
8	Shape and space	Unit 2: Braille and beyond	Reasoning	40–43
		Unit 3: Fenced in	Enquiry	50–53
		Unit 4: Tiling to order	Creative thinking	56–59
9	Measures	Unit 1: Easy as pi	Information processing	26–29
		Unit 3: Fenced in	Enquiry	50–53
		Unit 6: Classroom reshuffle	Using and applying	80–83
10	Measures including problems	Unit 1: Easy as pi	Information processing	26–29
11	Addition and subtraction	Unit 2: Chains	Reasoning	36–39
		Unit 4: At sixes and sevens	Creative thinking	60–63
		Unit 5: Domino dots	Evaluation	70–73
12	Properties of number	Unit 2: Chains	Reasoning	36–39
13	**Assess and review**			

Spring Term				
		Thinking by Numbers		
Unit	**Unit topic**	**Activity name**	**Thinking skill**	**Page no.**
1	Place value	Unit 1: Who wants to be a millionaire?	Information processing	30–33
2	Problem solving	Unit 1: Who wants to be a millionaire?	Information processing	30–33
		Unit 1: Easy as pi	Information processing	26–29
		Unit 2: Braille and beyond	Reasoning	40–43
		Unit 6: What's the issue?	Using and applying	84–87
3	Multiplication and division	Unit 3: Lost in the loop	Enquiry	46–49
		Unit 4: At sixes and sevens	Creative thinking	60–63
		Unit 5: Domino dots	Evaluation	70–73
4	Fractions and decimals	Unit 5: So what?	Evaluation	66–69
		Unit 6: Classroom reshuffle	Using and applying	80–83
5a	Shape and space	Unit 2: Braille and beyond	Reasoning	40–43
		Unit 3: Fenced in	Enquiry	50–53
		Unit 4: Tiling to order	Creative thinking	56–59
5b	Angle			
6	**Assess and review**			
7	Measures	Unit 1: Easy as pi	Information processing	26–29
		Unit 3: Fenced in	Enquiry	50–53
		Unit 6: Classroom reshuffle	Using and applying	80–83
8	Handling data	Unit 5: So what?	Evaluation	66–69
		Unit 6: What's the issue?	Using and applying	84–87
9	Addition and subtraction 1	Unit 2: Chains	Reasoning	36–39
		Unit 4: At sixes and sevens	Creative thinking	60–63
		Unit 5: Domino dots	Evaluation	70–73
10	Addition and subtraction 2	Unit 2: Chains	Reasoning	36–39
		Unit 4: At sixes and sevens	Creative thinking	60–63
		Unit 5: Domino dots	Evaluation	70–73
11	Reasoning about numbers			
12	**Assess and review**			

Summer Term

Unit	Unit topic	Thinking by Numbers		
		Activity name	Thinking skill	Page no.
1	Place value	Unit 1: Who wants to be a millionaire?	Information processing	30–33
2	Multiplication and division 1	Unit 3: Lost in the loop	Enquiry	46–49
		Unit 4: At sixes and sevens	Creative thinking	60–63
		Unit 5: Domino dots	Evaluation	70–73
3	Multiplication and division 2	Unit 3: Lost in the loop	Enquiry	46–49
		Unit 4: At sixes and sevens	Creative thinking	60–63
		Unit 5: Domino dots	Evaluation	70–73
4	Fractions, decimals and percentages	Unit 5: So what?	Evaluation	66–69
		Unit 6: Classroom reshuffle	Using and applying	80–83
5	Fractions, decimals, percentages, ratio and proportion	Unit 5: So what?	Evaluation	66–69
		Unit 6: Classroom reshuffle	Using and applying	80–83
6a	Handling data	Unit 5: So what?	Evaluation	66–69
		Unit 6: What's the issue?	Using and applying	84–87
6b	Solving problems and puzzles	Unit 1: Who wants to be a millionaire?	Information processing	30–33
		Unit 1: Easy as pi	Information processing	26–29
		Unit 2: Braille and beyond	Reasoning	40–43
		Unit 6: What's the issue?	Using and applying	84–87
7	**Assess and review**			
8	Shape and Space	Unit 2: Braille and beyond	Reasoning	40–43
		Unit 3: Fenced in	Enquiry	50–53
		Unit 4: Tiling to order	Creative thinking	56–59
9	Shape, space and measures	Unit 1: Easy as pi	Information processing	26–29
		Unit 2: Braille and beyond	Reasoning	40–43
		Unit 3: Fenced in	Enquiry	50–53
		Unit 4: Tiling to order	Creative thinking	56–59
		Unit 6: Classroom reshuffle	Using and applying	80–83
10	Measures and problem solving	Unit 1: Easy as pi	Information processing	26–29
11	Addition and subtraction	Unit 2: Chains	Reasoning	36–39
		Unit 4: At sixes and sevens	Creative thinking	60–63
		Unit 5: Domino dots	Evaluation	70–73
12	Properties of numbers and number sequences	Unit 2: Chains	Reasoning	36–39
13	**Assess and review**			

Thinking by Numbers 5 and the NNS Framework

Thinking skill	Unit	Activity name	Place value, ordering and rounding	Properties of numbers and number sequences	Fractions, decimals and percentages, ratio and proportion	Rapid recall of addition and subtraction facts	Mental calculation strategies (+ and −)	Paper and pencil procedures (+ and −)	Understanding multiplication and division	Rapid recall of multiplication and division facts	Mental calculation strategies (× and ÷)	Paper and pencil procedures (× and ÷)	Using a calculator	Checking results of calculations	Making decisions	Reasoning and generalising about numbers or shapes	Problems involving 'real life', money or measures	Measures	Shape and space	Organizing and interpreting data
Information processing	1	Easy as pi	✓														✓	✓		
		Who wants to be a millionaire?															✓			
Reasoning	2	Chains		✓		✓													✓	
		Braille and beyond														✓				
Enquiry	3	Lost in the loop								✓	✓							✓	✓	
		Fenced in																	✓	
Creative thinking	4	Tiling to order				✓			✓											
		At sixes and sevens			✓															
Evaluation	5	So what?					✓				✓			✓						✓
		Domino dots			✓															
Using and applying thinking skills	6	Game on																		
		Classroom reshuffle																✓		
		What's the issue?															✓			✓

Thinking by Numbers 5 and the 5–14 Guidelines

Thinking skill	Unit	Activity name	Problem solving and Enquiry	Information Handling	Range and Type of Numbers	Money	Add and Subtract	Multiply and Divide	Round Numbers	Fractions, Percentages and Ratio	Patterns and Sequences	Functions and Equations	Measure and Estimate	Time	Perimeter, Formulae and Scales	Shape, Position and Movement
Information processing	1	Easy as pi	✓										✓			
Information processing	1	Who wants to be a millionaire?	✓			✓			✓				✓			
Reasoning	2	Chains			✓		✓									
Reasoning	2	Braille and beyond						✓								✓
Enquiry	3	Lost in the loop						✓								
Enquiry	3	Fenced in											✓			
Creative thinking	4	Tiling to order					✓	✓								✓
Creative thinking	4	At sixes and sevens								✓						✓
Evaluation	5	So what?		✓												
Evaluation	5	Domino dots					✓	✓								
Using and applying thinking skills	6	Game on	✓													
Using and applying thinking skills	6	Classroom reshuffle								✓			✓			
Using and applying thinking skills	6	What's the issue?	✓	✓												

Glossary

algorithm a step by step procedure that, if followed exactly, will always yield a correct solution to a type of problem

assessment for learning an approach to **formative assessment** where the learner is encouraged to take responsibility for evaluating their own achievement of learning objectives. An aspect of **self-regulation**.

Bloom's Taxonomy a widely used instructional objectives model developed by the prominent educator Benjamin Bloom and colleagues in the 1950s. It categorizes the cognitive, affective and conative domains and includes a systematic list of thinking skills, in categories and sub-categories such as comprehension, application, analysis, synthesis, and evaluation. The last three are considered **higher-order** thinking skills.

brain-based learning a range of techniques and approaches to teaching and learning which take their inspiration from research into how the brain works to identify implications for teaching

brainstorm a technique for rapid production of ideas without critical examination, evaluation or elaboration

bridging a teaching strategy where explicit links are drawn from what has been learned to other related contexts to help **transfer**

cognition the mental operations involved in thinking; the biological/neurological processes of the brain that facilitate thought. Sometimes contrasted with affect or emotion and conation (wanting or willing).

Community of Enquiry the process of developing knowledge and understanding by participating in purposeful dialogue or collaborative discussion. Also the teaching technique used in Philosophy for Children with a class of pupils.

concrete preparation an introductory phase in some teaching thinking approaches where new words are introduced and learners become familiar with what the task is about

constructivism a view of learning in which learners are seen as building or developing their own understanding of how the world works from their experience and interaction with people around them

creative thinking producing new ideas or thoughts. Imaginative thinking that is aimed at producing outcomes that involve synthesis of ideas or lateral thinking; thinking that is not analytical or deductive, sometimes referred to as divergent thinking.

critical thinking a generic term for thinking skills used in the United States. The process of determining the authenticity, accuracy, or value of something; characterized by the ability to seek reasons and alternatives, perceive the complete situation, and change one's view based on evidence and reasoning. Sometimes also called analytical or convergent thinking. Often related to formal or informal logic and to reasoning.

demonstrating showing children how to do something, how to perform a skill or a technique, how to carry out a process, how to repeat and practise what they have been shown

dialogue shared enquiry between two or more people

enquiry a systematic or scientific process for answering questions and solving problems based on gathering evidence through observation, analysis and reflection

enquiry learning a teaching strategy designed to develop pupils learning through systematic gathering of observation and investigation

enrichment an approach to teaching thinking as separate discrete skills, usually as separate lessons using a particular programme or set of activities

formative assessment assessment which alters subsequent teaching and learning. This may involve teachers in using information gathered in lessons to alter what they do (see **mediation**) or it may also involve the learner through **assessment for learning**.

graphic organizers diagrams which help learners to organize information such as by comparing and contrasting using a grid of similarities and differences

heuristics general or widely applicable problem-solving strategies. Guidelines that generally direct attention, but that do not always produce a correct outcome (see **algorithm**).

higher order thinking evaluation, synthesis and analysis, the higher levels of **Bloom's Taxonomy**

infusion integrating thinking skills teaching into the regular curriculum or lessons; infused programs are commonly contrasted with **enrichment** programs, where separate or discrete skills are taught through lessons to promote thinking.

mediation a teaching strategy where the teacher intervenes and supports the development of pupils' understanding by **modelling** or by direct instruction to help them achieve something they could not do alone

metacognition the process of planning, assessing, and monitoring one's own thinking. Thinking about thinking in order to develop understanding or **self-regulation**.

modelling teaching children in a way that helps them to see the underlying structures, and to understand the embedded or supporting concepts and ideas

multiple intelligences the idea developed by Howard Gardner that IQ does not measure aspects of intelligence sufficiently and that people have strengths in different areas such as visual-spatial or musical as well as more traditionally assessed areas such as linguistic or logico-mathematical

problem based learning an approach using **problem solving** techniques where learners are set specific challenges through realistic or unstructured problems. Similar to **enquiry learning**, but with a particular goal or challenge which needs to be resolved

problem solving a general term which covers a diversity of problem types which make a range of demands on thinking. Some problems have unique solutions and can be tackled with predominantly convergent critical thinking, but many others are open-ended and demand both creative and critical thinking for their solution.

reasoning drawing conclusions or inferences from observations, facts, experiences: deductive inferring conclusions from premises; inductive: inferring a provisional conclusion or hypothesis from information

self-regulation the conscious use of mental strategies to improve thinking and learning, often aimed at particular learning goals

seriation sequencing or arranging objects, ideas or events in a particular order determined by a criterion

Socratic questioning an approach to questioning and discussion where answers to questions are pursued through dialogue

thinking skills 'thinking skills' and related terms are used to indicate a teaching approach which emphasizes the processes of thinking and learning that can be used in a range of contexts. The list of thinking skills in the English National Curriculum is similar to many such lists: information-processing, reasoning, enquiry, creative thinking and evaluation.

transfer the ability to apply an idea or a skill that has been learnt in one context and use it in a different context